PREFACE

World-wide activity in the area of adiabatic diesel engine technology continues to accelerate. Taken from the Adiabatic Engine Session of the 1984 SAE International Congress and Exposition in Detroit, this volume of papers represents diverse cross sections of international research and development activities in this general subject area, with a broad range of key technological areas covered. Papers from Asia, Europe, and the United States combine to produce a true international flavor. In addition, this volume includes initial discussions of work being performed on adiabatic engine/passenger car applications.

As technology advances continue to materialize, it becomes increasingly obvious that an evolution toward "minimum-cooled" adiabatic concept engines is inevitable. Recent advances in key engineering areas such as high temperature materials, high temperature tribology, performance modeling, and advanced ceramic engine design have combined to provide a firm path to this evolution. The concept centers around insulating as much of the combustion chamber as possible, thereby reducing or even eliminating, the heat rejection to the coolant. The resulting high energy exhaust gases may be utilized in an external system such as turbocompounding. Potential advantages of such an engine are both numerous and significant. They include virtual elimination of the conventional cooling system, dramatic improvements in fuel economy, reduction in engine size and weight, improvement in reliability and maintainability, improved multifuel characteristics, smoother combustion, less noise, improved cold start and emission characteristics, and reduced manufacturing costs.

Future worldwide, aggressive efforts in the area of adiabatic engine technology have been assured, with their application limited only by the imagination of man.

Walter Bryzik
Co-Chairman and Organizer
US Army Tank-Automotive Command

Roy Kamo
Co-Chairman and Organizer
Cummins Engine Co.

ADIABATIC ENGINES:

WORLDWIDE REVIEW

SP-571

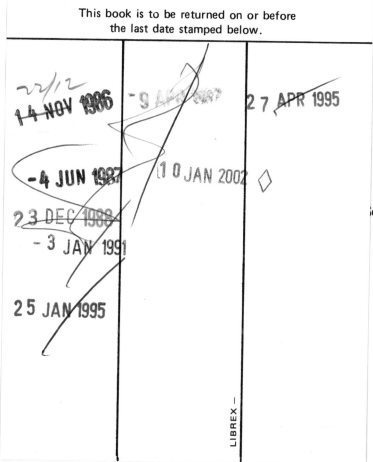

This book is to be returned on or before
the last date stamped below.

Published by:
Society of Automotive Engineers, Inc.
400 Commonwealth Drive
Warrendale, PA 15096
February 1984

ISBN 0-89883-342-6
SAE/SP-84/571
Library of Congress Catalog Card Number: 83-51756
Copyright 1984 Society of Automotive Engineers, Inc.

TABLE OF CONTENTS

Development of Ceramic Pre-Combustion Chamber for the Automotive Diesel Engine

H. Matsuoka and H. Kawamura
Isuzu Motors Ltd.
S. Toeda
Kyocera Ltd.

ABSTRACT

A pre-combustion chamber (hot plug) made of the silicon nitride ceramics has been developed. The hot plug constitutes a component of the combustion chamber of swirl chamber diesel engines. And it is subjected to the severest thermal load of all the diesel engine operating components. Unlike metal hot plugs, the ceramic hot plug application requires a unique design approach to meet the ability of the ceramic materials. That is, the soft engine mounting is necessary to avoid the concentraction of mechanical stress resulting from the high Young's Modulus of the ceramic and control of temperature distribution is also required to reduce thermal stress.

Due to the heat insulating construction, the ceramic hot plug permits the combustion performance to be improved at a low speed and load, resulting in improved noise, startability, and HC emission.

In additional the ceramic hot plugs are already being produced for diesel passenger car.

DURING THE LAST FEW YEARS, there is an increasing demand to reduce the automobile fuel consumption. In response to this necessity, together with the positive efforts for effective body weight reduction, improvement in aero-dynamic features, and lowering of friction, a wide range of detailed research projects are being carried out to make engines more fuel-efficient.

As a structural material, ceramic has better corrosion resistance, high-temperature strength, and hardness than metals, and in comparison with plastics, not withstanding the light weight, has superior hardness and precise moldability. Furthermore, with the outstanding development in its high-temperature strength, heat shock resistance, and wear resistance, the ceramic materials have become ideally suitable for diesel engines which are drawing attention for their low fuel consumption and use of alternative type of fuel.

In this paper we will discuss the research being carried out on long-life ceramic hot plugs in succession to the ceramic glow plugs which are similarly exposed to high-temperature gases in the combustion chambers.

DESIGN CONCEPT OF CERAMIC MATERIALS

The development of ceramic materials, ranging from the material containing metallic oxide containing nonmetallic oxide as the main component has increased the possibility of development of engineering ceramics. In case where the ceramic material is used in the same way as metallic material, many troubles will be created, making it difficult to utilize to the fullest its excellent high temperature strength. This paper describes improved attaching method and shape of hot plug in diesel engine prechamber tending to be exposed to the severest temperature of the engine in order to replace metallic hot plug with ceramic one and the need for a wide range of studies on mechanical and high temperature strength of ceramic material required for volume production.

Fig.1 shows a general view of prechamber and piston commonly used in small diesel engine. The flange of hot plug is press fitted in cylinder head partly with prechamber wall surface on the cylinder head side exposed to main combustion chamber and partly contacting cylinder block with gasket interposed. In such configuration, however, mechanical stress is localized on stepped portion of the flange. Fig.2 indicates the stress concentration obtained by finite element analysis. A hot plug which is made of the ceramic material having almost the same mechanical strength as the metallic hot plug and fitted into engine in the same shape invariably breaks. Even if ceramic material strength is brought up to such a level as to withstand the peak localized stress concentration, the probability of the hot plug breakage is extremely high. That is, the ceramics is more sensitive to the localized stress concentration than metal and

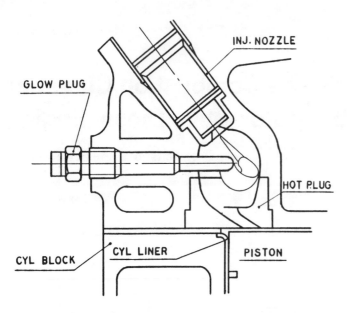

Fig. 1   Schematic of Pre-Combustion Chamber

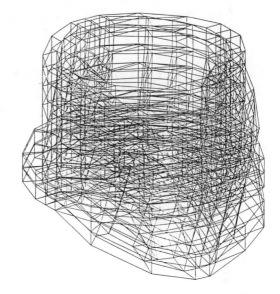

Fig. 2   Static Analysis Using Finite Element Analysis

the shape and construction of ceramic hot plug allowing stress concentration in engine installed condition significantly increase the chance of the breakage. Such experience indicates the need for stresses working on the ceramic hot plug when installing the hot plug in engine to be distributed uniformly in all directions.

The ceramic engine components must be designed, giving considerations to the mechanical characteristics different from the metallic materials. Although the machanical stress can be reduced by the above methods, the ceramic hot plug cannot be assured of sufficient durability. It is highly sensitive to the thermal stress resulting mainly from temperature gradient and is very weak to abrupt heat shock.

It is, therefore, necessary to minimize mechanical and thermal stress concentration to bring out the excellent ceramics features.

STRENGTH OF CERAMIC HOT PLUGS

The hot plug is a constituent part of the pre-chamber, forming the connecting passage between the pre-combustion chamber and the main combustion chamber of the swirl chamber diesel engine, and is subjected to extremely high temperature as compressed air and combustion flame passes through this connecting passage. Fig. 3 shows the construction of the metal hot plugs which are currently in use.

Generally, the portion of the metal hot plug forming the constituent part of the combustion chamber acts as a heat insulator and is separated from the cylinder head by a very thin air layer. It has a collar for pressure fitting on the cylinder head. Being exposed to such rigorous thermal conditions, the hot plugs need to be carefully tested for their mechanical and thermal strength, and Fig.4 shows the shape of ceramic hot plug obtained with full considerations given to the mechanical and thermal strength of the

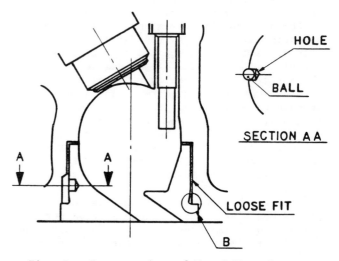

Fig. 3   Construction of Metal Hot plug

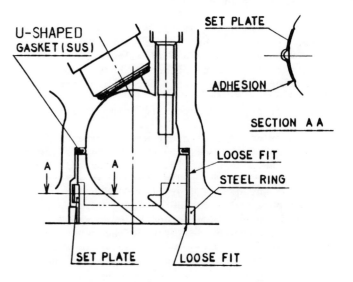

Fig. 4   Construction of Ceramic Hot Plug

plugs. That is, an elastic gasket is interposed between one end of the plug and the cylinder head and a very small clearance is provided between the outer periphery of hot plug and the cylinder head.

The configuration and construction of ceramic hot plug allowing for thermal expansion are determined by the mechanical and thermal strength of the ceramics. Therefore, control of mechanical stress on the plug when installed in the cylinder head is necessary.

MECHANICAL STRENGTH - As already noted, when metal hot plugs are replaced by identically shaped ceramic hot plugs (Si_3N_4), they are largely damaged due to stress concentration in the stepped parts "B" in Fig. 3. However, the flexure strength of ceramic material is 490 to 686 MPa which is the same as that for the metals (nickel chromium alloy) and it is difficult to think that these can be broken by stress concentration.

Deformation of the hot plug which is pressed in between the two large parts like the cylinder body and the cylinder head is controlled by the deformation of the cylinder head which is fixed with the cylinder body by large clamping force. Consequently, similar deformations are found in both metal and ceramic hot plugs. Since Young's modulus of Si_3N_4 is 1.5 times higher than the nickel chrome alloy, its maximum stress value is also 1.5 times higher at same deformations.

As shown in Fig.4, U-shaped gasket is placed between the upper end of the hot plug and the cylinder head and the lower flange is positioned by a steel collar. The hot plug is clearance fitted into the cylinder head to minimize the mechanical stress working on the plug when installed.

On the other hand, the fracture of the ceramics are largely dependent on their internal flaw, which can be expressed by the following equation.

$$\sigma = \sqrt{2Er_s/\pi C} \quad \ldots\ldots\ldots\ldots (1)$$

  E ; Young's Modulus
  r_s; surface energy of
       the flaw
  C ; half size of the
       flaw
  σ ; Stress

The fracture stress of ceramics obtained from Eq. (1) above is largely susceptible to the size of the internal flaw. When the value of the stress imposed upon the internal flaw approaches the valve caluculated in Eq. (1) above, the fracture is sudden. It is estimated that in ceramics the sensitivity to fractures generated by the intenal flaw size is approximately 10 times that of the metals. Another important feature of ceramics is the speed of its fracture. Hence, once the fracture starts it progesses rather rapidly. In the following equation the relation between the fracture speed and the internal flaw size shown.

$$V = A \cdot K_1^N \quad \ldots\ldots\ldots\ldots (2)$$

Where A and N are constant depending on the ceramic type, and K_1 the fracture toughness. K_1 can be expressed by the equation as follow.

$$K_1 = Y \cdot \sigma \cdot a^{1/2} \quad \ldots\ldots\ldots (3)$$

Where Y is a coefficient usually having a value between 1 to 2 depending on the ceramic shape and loading conditions. The normal value of N is between 20 and 50, and for silicon nitride, the value is between 40 to 50. Consequently, according Eq. (3), under certain operating conditions, the fracture speed is greatly influenced by the flaw size. However, as ceramics are made of fused particles, the internal flaws are common. Therefore, it is necessary to reduce the internal flaws of ceramics to a minimum and control its flaw size below $50\,\mu m$ caused by a value of stress.

Along with the control of the flaw size , it is also necessary to reduce the installation stresses as far as possible. As shown in Fig. 4, during installation of the ceramic hot plug, an 'U' shaped gasket is used on its upper end, and is fixed by a steel ring on the stepped parts. Mechanical stress during installation is reduced to a minimum by fitting the hot plugs to the cylinder head with gaps.

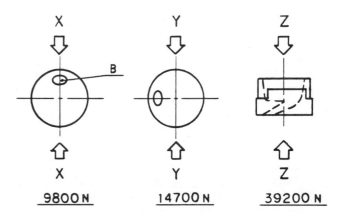

Fig. 5 Breaking Strength of Hot Plug

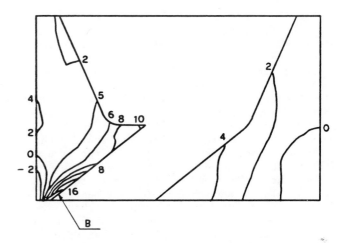

Fig. 6   Result of Finite Element Analysis

The reason for making that external shapes simple for ceramic hot plugs is to reduce stress concentration not only due to internal flaws but also the stepped external shape where stress

Fig. 7   Heat Cycle Test of Hot Plug

concentration is much higher than metal hot plugs. Fig. 5 shows the test result of breaking strength with static loads from various directions to investigate the non-uniformity strength of ceramics. Breaking strength of the X-X direction shows the lowest value caused by tensile stress at the internal side of the throat of the hot plug. The maximum stress during this period is 441 MPa and since the hot plug is very weak in this direction, it is necessary to reduce excessive stress caused by such external factors as stated before. Figure 6 shows the results of finite element analysis which agree with actual measured values.

THERMAL LOAD STRENGTH – When ceramic materials are used as engine components, apart from structural stresses it is also necessary to alleviate their thermal stresses. Regarding the heat resisting strength, the study should be based on the following 3 subjects: the thermal stress fatigue caused by the heat cycle test; change in surface composition under high temperature oxidizing atmosphere; checking the thermal shock by dipping the high temperature parts in water. To check these special characteristics, test piece measurements are insufficient, and it is necessary to carry out tests and measurements with actual components.

Results of thermal stress fatigue caused by heat cycle is shown in Fig. 7. A rotary equipment provided with a gas burner and a cooling air nozzle for alternately heating and cooling the surface is used especially to check the strength loss at the sharp edge of the throat outlet inside the combustion chamber due to the heat cycles. The temperature cycle is repeated by raising it to 750°C in 2 minutes and then lowering to 100°C in 3 minutes. On checking, no significant loss in strength was found after completion of these tests. Similar tests on nickel chromium hot plugs showed progressive oxidized surface corrosion.

Fig. 8 shows the result of strength measurement of hot plugs kept for 1 hour in a high temperature oxidized atmosphere. Silicon nitrides undergo surface composition alteration

when exposed to high temperatures. This layer of altered composition, similar to cracks is very brittle in nature. Consequently, these surfaces become very weak and start to fracture when subjected to tensions. Although the result shown here is for 1 hour test in high temperature, but in fact, it is found that prolonging exposure has no adverse influence on the strength. Hence it is concluded that the ceramic's fracture strength is dependent more on temperature level rather than the exposure time.

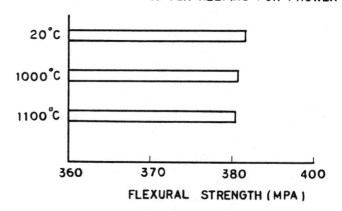

Fig. 8   Exposure Test in High Temperature Oxidized Atmosphere

Even after carrying out sufficient unit tests to confirm the mechanical and heat resisting strengths, cracks do appear on hot plugs subjected to endurance tests on actual engines. Therefore, for hot plugs there is a wide difference in temperature distribution and temperature gradient between actual operation in an engine and a unit test. Hot plugs are excessively affected by thermal stresses.

In Fig. 9, temperature measurements carried out with templugs in the major areas of the hot plug during operation is shown. It can be noted that the combustion chamber side of the hot plug has high temperature, whereas, on the gasket side it is lower by 200°C. Furthermore, the combustion chamber side of the hot plug is exposed

4

to hot gases having a temperature between 2000 to 2500°C, and in spite of an internal temperature rise of about 900°C in that area, the surface in contact with the cylinder head side shows a temperature of only 300°C. Hence, the temperature gradient is remarkably large. The following considerations are given for the hot plug's thermal stresses on the basis of temperature measurements made on this side.

In the ceramic hot plugs having high Young's modulus, the temperature distribution causes compressive stress on the high temperature parts of the internal surface, tensile stress on the low temperature parts of the external surface. Further, the combustion chamber side of the hot plug is subjected to compressive stress, and the internal surface of the cylinder head side (portion marked S in Fig. 10) to rather large tensile stress.

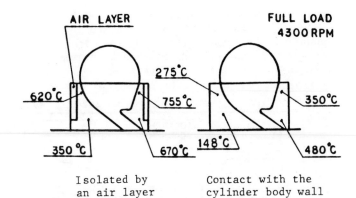

Fig. 11   Effect of Air Layer at Outer Surrounding on Hot Plug Temperature

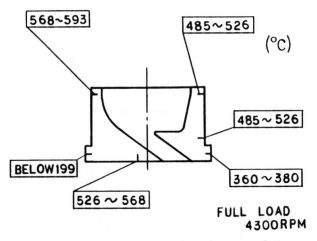

Fig. 9   Temperature Distribution of Hot Plug

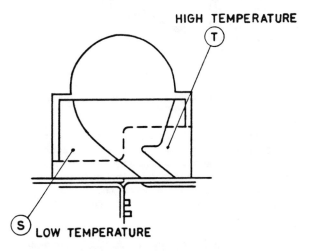

Fig. 12   Thermal Control of Hot Plug

Consequently, to reduce thermal stresses of hot plugs, the temperature distribution should be uniform as far as possible. In Fig. 11, the temperature distribution in the external surface of the hot plug isolated by an air layer is compared with the temperature distribution of the low temperature surface in contact with the wall surface of cylinder body. In comparison with the former, the temperature of the latter is 300°C lower. Consequently, temperature control of hot plug is possible by reducing the wall surface contact area. According to this method, the temperature difference between the above mentioned S and T portion can be reduced by largely increasing the internal contact surface area (T portion) of combustion chamber side and reducing the internal contact surface area (S portion) of the cylinder head gasket side. In Fig. 13, the temperature distribution in the temperature controlled type hot plug shown in Fig. 12 is compared with the temperature distribution of the existing types. With this method, thermal stress reduction in hot plugs is possible but in cases like Fig. 12, where the portion T is in contact with the cylinder head, the thermal stress is excessively affected by temperature gradient existing in the small confined area between the high temperature portion of the gas passage and

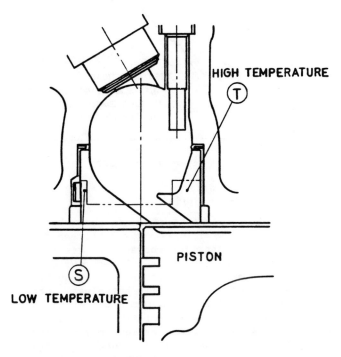

Fig. 10   Thermal Stress of Hot Plug

low temperature portion of the cylinder head. Moreover, due to smaller thickness and low mechanical strength of the T portion, it is necessary to control and maintain high temperature in this part to generate compressive stress.

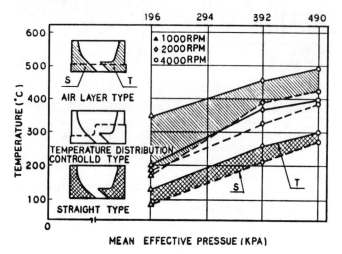

Fig. 13 Comparison of Hot Plug Temperature Distribution

With these heat control and thermal stress control methods it became possible to reduce the thermal stresses generated inside the combustion chamber due to the sudden heating and cooling cycles, and in combination with the mechanical stress reduction method stated in this clause we completed a remarkably long-life ceramic hot plug.

The locating holes shown in Fig.3 considered as subject to tensile stress from the view point of thermal stress and as the starting point of stress concentration sensitive to the mechanical stress. Therefore, the positioning method was changed to metal plate method shown in Fig.4.

PERFORMANCE

Ceramic hot plugs are highly reliable and durable in high-temperature combustion gases due to reduction in mechanical and thermal stresses. Accordingly, by utilizing this thermal strength of ceramics in swirl-chambered diesel engines, the performance can be improved.

In swirl-chambered diesel engines, since the fuel jet sprayed in the high-speed vortex changes into stratified air-fuel mixture, the mixture zone at the outer area of the vortex is remarkably influenced by the combustion chamber temperature. Therefore, it is better to have high temperature in the hot plug which forms a part of the combustion chamber. This effect appears especially during low speed and low load conditions.

In Fig. 14, temperatures at various parts of the combustion chamber hot plugs of similar type made respectively of ceramic and metal are compared, and the ceramic hot plug maintain a higher temperature (between 30 to 60°C). Since ceramic hot plug temperatures can be controlled

by increasing or decreasing the conduction area, it is possible to achieve better combustion during low load conditions. In small diesel engines, specially during starting and idling when the combustion chamber wall temperature is low, the ignition of the atomized fuel is delayed causing emission of unburnt gases in the form of blue-white smoke, and the noise tends to increase due to the increase in rate of pressure rise caused by the increase in pre-mixture combustion.

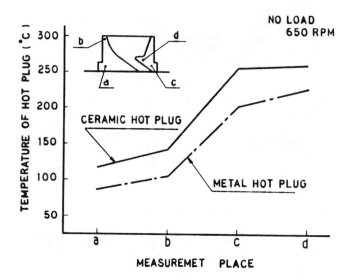

Fig. 14 Temperature Comparison at Various Parts

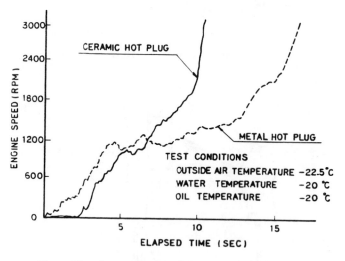

Fig. 15 Comparison of Low-temperature Startability

Fig. 15 shows the comparison between the low-temperature starting characteristics of ceramic and metal hot plugs. Low-temperature starting characteristics of an engine can be improved by using the ceramic hot plugs since the heat capacity per unit volume of ceramic hot plug is approximately 1/2 of the metal hot plugs which allows faster temperature rise, and its heat-insulating construction reduces heat radiation from the swirl chamber walls and thus increases the temperature of the compressed air.

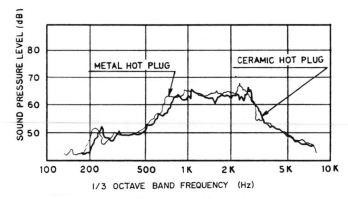

Fig. 16 Noise Comparison during Idling at
Low Temperature

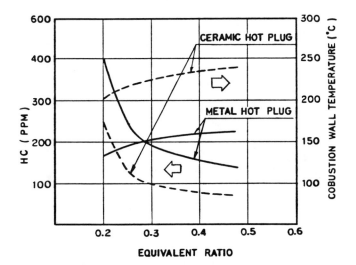

Fig. 17 Combustion Chamber Temperature
and HC

From Fig. 16, which shows the results of noise frequency analysis results idling with 30°C water temperature after low-temperature starting, it can be noticed that sound pressure levels by using ceramic hot plug is large reduction at high frequency area, and an overall reduction of 1 to 2dB has been measured.

Figure 17 shows HC emission against combustion wall temperature during low speed, low load operations. Ceramic hot plug had about 70°C higher combustion temperature and 50 to 70 ppm lesser HC emission compared to the metal hot plugs.

CONCLUSIONS

Among all the operating parts of an internal combustion engine, the hot plug of the swirl-chambered diesel engine is subjected to the most severe thermal load. With the use of ceramic materials for the hot plugs the following points have become clear.

(1) Silicon nitride ceramics have sufficient resistance against oxidizing high temperature combustion gases of diesel engines and can be used as a construction material for parts exposed to high temperature.

(2) When using silicon nitride ceramics as construction material for high temperature parts of internal combustion engines, a widely different design technique than that of the metals should be employed using the following special characteristics of the ceramic materials:
(a) As compared to steels, ceramics have 1.5 to 2 higher Young's modulus. Regarding the deformation of other constituent parts caused by thermal and mechanical stress, ceramics area is easily susceptible to stress concentrations from their contact with the affected parts since their stress dispersion is less depending on the change in form. Therefore, the contact surface with the metal parts needs to be soft mounted with an elastic material.

(b) To reduce thermal stress caused by extremely high sectional temperature gradients and receiving large heat cycle, the heat conducting surface must have sufficient air layers and control the area of air layers.

(3) Thermal insulation construction of hot plugs reduces noise level, improves starting characteristics and improvement of HC emission by improving combustion during low speed, low load operation.

(4) To use ceramics as construction materials for high-temperature parts, the existence of internal flaws must be minimized as far as possible. To accomplish this manufacturing and sintering processes should be controlled, and the establishment of the non-destructive inspection method is important.

REFERENCES

(1) A.A. Griffith, Phil. Trans. Roy, Soc., 221 (1920)
(2) A.G. Evans, "Fracture Mechanics of Ceramics" (1976)

840427

Investigation on the Differentially Cooled System

C. S. Kirloskar, S. B. Chandorkar
and N. N. Narayan Rao
Kirloskar Oil Engines Limited
Pune, India

ABSTRACT

In order to clarify certain aspects of the combustion process in differentially cooled engines, a special engine was built in which the surface temperatures of the piston crown, cylinder liner and cylinder head bottom could be varied independently. A computer programme to calculate the swirl and turbulence intensity in the combustion chamber was developed. Investigations showed that in order to improve the mixing process, the liner temperature should be kept low while the piston temperature may be increased only if the cylinder head temperature is kept low, and vice versa. A low coefficient of regression in the plotted results indicated that the flow parameters investigated were not the determining parameters but only one among many others which need further investigation. The present investigation however indicates that the high piston temperature found desirable earlier should preferably be combined with low cylinder head and liner temperatures.

THE DIFFERENTIALLY COOLED SYSTEM incorporates a cast iron piston working in a cylinder liner with reduced cooling, the cylinder head being conventionally cooled. When applied to a small naturally aspirated diesel engine, it had resulted in a higher thermal efficiency and smoke-limited BMEP [1]*. In a later paper [2], this had been ascribed to a more favourable rate of heat release diagram. It was clear that higher wall temperatures within certain limits were improving the combustion process in the engine. Since the combustion process is a complex phenomenon, it was

---

* Numbers in parentheses designate references at end of paper.

proposed to split it into several steps and to investigate the effect of higher wall temperatures on each step. This paper summarises the results so far obtained from investigations into the flow parameters of swirl, squish and turbulence. The results cannot be treated as conclusive but do indicate a few interesting trends in the pre-conditions necessary for efficient combustion.

For purposes of investigation a special engine was built in which the surface temperatures of the piston crown, cylinder liner and cylinder head bottom could be varied artificially. The results were correlated with calculations based on a swirl and turbulence intensity programme. The analytical and experimental techniques used are detailed below followed by discussion and conclusions.

ANALYTICAL TECHNIQUES

The important analytical tools used were a cycle simulation programme, and a swirl and turbulence intensity programme, with additional assistance derived from an axi-symmetric finite element programme and a few minor programmes.

CYCLE SIMULATION PROGAMME — This was based on the filling and emptying model for induction and exhaust processes combined with Watson's equations [3] for rate of heat release. Delay period according to Wolfer[4], heat transfer according to Woschni [5], internal energy according to Krieger and Borman [6], and frictional losses according to Millington and Hartley [7] were incorporated in the programme. The 4th order Runge-Kutta method of numerical solution of differential equations was used [8].

Watson's equations for rate of heat release consist of two parts: an empirical equation for the period of uncontrolled combustion:

$$F_1 = C_1 C_2 \, y^{C_1 - 1} \left(1 - y^{C_1}\right)^{C_2 - 1} \quad (1)$$

and a Weibe function for the period of diffusion burning :

$$F_2 = C_3 C_4 \, y^{C_4 - 1} \, exp\left(-C_3 \, y^{C_4}\right) \quad (2)$$

The rate of heat release is then the sum of the two components:

$$\frac{dQ_B}{d\phi} = \frac{Q_{BO}}{\phi_{VE} - \phi_{VA}} \left[G F_1 + (1 - G) F_2\right] \quad (3)$$

In these equations, y is the fractional relative time elapsed at the crankangle ϕ :

$$y = \left| \frac{\phi - \phi_{VA}}{\phi_{VE} - \phi_{VA}} \right| \quad (4)$$

and G is the fraction of fuel burnt almost instantaneously at the end of the delay period:

$$G = 1 - \frac{0.71}{\lambda_a^{0.37}} \left(\frac{n}{\Delta\phi_{zv}}\right)^{0.26} \quad (5)$$

where, n = engine speed (RPS)
$\Delta\phi_{ZV}$ = delay period, ($^\circ$crankangle)
λ_a = relative air-fuel ratio
ϕ_{VE} and ϕ_{VA} are the crankangles at end and beginning of combustion respectively. Q_B and Q_{BO} are the instantaneous and total quantity of heat released. C_1, C_2, C_3 and C_4 are constants given by :

$$C_1 = 2 + 2.6875 \times 10^{-3} \Delta\phi_{zv} \quad (6)$$

$$C_2 = 5000 \quad (7)$$

$$C_3 = 14.2 \, \lambda_a^{1.0} \quad (8)$$

$$C_4 = 1.125 \, C_3^{0.25} \quad (9)$$

SWIRL AND TURBULENCE PROGRAMME - Fig.1 illustrates the idealization used for the working space in the engine. Two control volumes, V_1 above the cavity, and V_2 above the piston crown deck, have a common boundary through which squish action takes place. The cavity was hemispherical and centrally located in the piston, as in the actual engine. The incoming angular momentum of the swirl was assumed to be divided in the ratio of the two

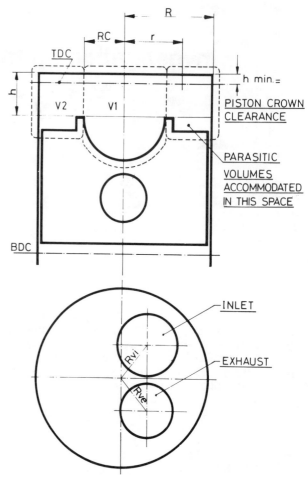

Fig.1. Idealization of engine working space

volumes V_1 and V_2. The swirl was treated as axi-symmetric and the tangential velocity distribution was assumed to be parabolic (Fig.2) according to the formula:

$$V_t = \alpha \, r + \beta \, r^2 \quad (10)$$

where the constants α and β were calculated by solving the differential equations for total angular momentum in each volume. From this, the squish velocity could also be derived. The turbulence intensity as well as the pressure, temperature and density were assumed constant throughout both the volumes. The $K - \epsilon$ model was used to calculate the turbulence intensity basically following the scheme and constants of Borgnakke [9] and Launder and Spalding [10] with the following three important modifications:

(a) A logarithmic velocity distribution was introduced for the turbulent boundary layer in equilibrium at the wall, as given by Schlichting [11].

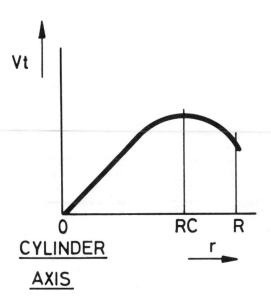

$$Vt = \alpha r + \beta r^2$$

$\alpha \, \& \, \beta$ ARE CALCULATED BY
THE PROGRAM

Fig.2. Assumed variation of tangential velocity with radius of cylinder

(b) The additional turbulence intensity created by the squish component was added.
(c) First order Euler's integration procedure was used to solve the differential equations. The step width was so selected that the change of any dependent variable per step was not more than 5% of its instantaneous value.

The programme was validated by comparison with the published results on the PROCCO engine [12]. A more detailed description of the programme will be published separately.

The programme is capable of simulating a motored engine and provides the following results:

(a) Turbulent velocity $\quad u' = \sqrt{2K/3} \quad$ (11)

Where K is the turbulent kinetic energy.

(b) Turbulence mixing time $\lambda_t = \sqrt{\nu/\epsilon}$ (12)
Where ν is kinematic viscosity of air
ϵ is the rate of dissipation of K

(c) Turbulence mixing length $\lambda = u'\sqrt{\dfrac{15\nu}{\epsilon}}$ (13)

(d) Turbulent kinematic viscosity $\qquad$ (14)
$$\nu_t = C_D K^2/\epsilon$$

where C_D is a constant equal to 0.09 [10]
(e) Swirl ratio in the cup

$$S = \frac{\displaystyle\int_0^R V_t \, r \, dm}{2\pi n \displaystyle\int_0^R r^2 \, dm} \qquad (15)$$

where n is the engine speed (RPS) and dm is the mass of an element of air. Definitions (12) and (13) are from [13].

The introduction of the actual combustion process in the programme was avoided on the assumption that the flow conditions just before the beginning of combustion are not altered significantly whether the engine is motored or fired. In order to make the results realistic, the cylinder conditions at the beginning of calculation ($\equiv$ Crankangle at exhaust valve opening) were assumed to be equal to those obtained with the cycle simulation programme at the same point, and each of the calculations was only allowed to run once without iteration.

FINITE ELEMENT PROGRAMME - A programme developed by Wilson [14] for axi-symmetric solids was used to calculate the thermal

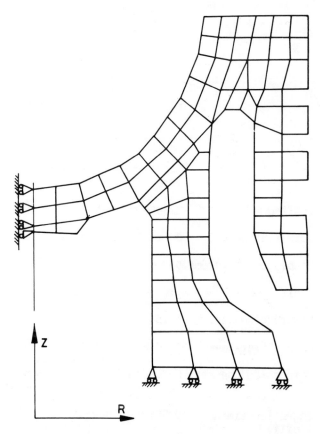

Fig. 3. Model for finite element analysis of piston showing the assumed boundary conditions

expansion of the cast iron piston. The piston was idealised using axi-symmetric ring elements of quadrilateral cross section. As the calculation was done for TDC position, inertia and side thrust forces were ignored. The material was treated as being elastic with properties independent of temperature which may be considered valid for the range of temperatures investigated. The material properties used were; modulus of elasticity $E = 1.18 \times 10^8$ kPa, poisson's ratio = 0.3 and coefficient of thermal expansion = $11 \times 10^{-6}/{}^\circ K$.

A radial section of the piston through the gudgeon pin was modelled with 125 nodes and 86 elements. Peak firing pressure and temperature distribution in the piston were inputs. The boundary conditions assumed are shown in Fig.3.

EXPERIMENTAL TECHNIQUES

Three techniques were used for the experimental part of the work namely, engine tests on a dynamometer, an Electrical Analog for piston temperature distribution, and a swirl test rig for definition of inlet flow.

ENGINE TESTS - The engine used for the test had the standard specifications shown in Table I.

Table - 1 Engine Specifications

Item	Specifications
Type	Single cylinder, 4-stroke vertical, diesel
Cooling	Liquid cooled
Bore x Stroke	80 mm Ø x 110 mm
Displacement volume	553 cm^3
Aspiration	Naturally aspirated
Compression Ratio	16.5:1
Rated Power/Speed	3.67 kW/1500 rpm
Combustion Chamber	D.I. with centrally located hemispherical cup in piston
Injection timing (Spill)	27°BTDC
Inlet Port	Lip-in-port type

The engine was modified in the following respects:
(1) The aluminium alloy piston was replaced with a cast iron piston which is normally used in the differentially cooled version of the same engine.
(2) The cylinder water jacket and cylinder head water jacket were isolated from each other and each was provided with its own separate closed coolant circuit. Thus the cylinder liner and cylinder head could be cooled independently to different levels. Ethylene Glycol was used as coolant to permit wider variation of surface temperatures than are possible with water.
(3) The connecting rod was provided with a drilled hole through which lubricating oil could be supplied to cool the piston undercrown. A screw was fitted across the drilled hole near the big end of the connecting rod by means of which the drilled hole could be closed, opened, or opened partially to control the extent of piston cooling, and thereby to vary the piston crown surface temperature.
(4) Lubricating oil of MIL-L-2104C grade of detergency was used to minimise ring sticking or other problems in the course of the experiments.

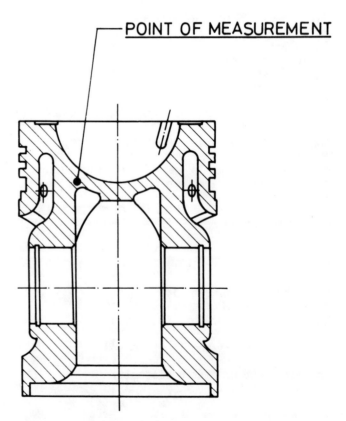

POINT OF MEASUREMENT

Fig. 4. Point of measurement of temperature in the piston

Although these modifications enabled the surface temperatures of the piston crown, liner and cylinder head to be independently varied in principle, the interaction between the three temperatures could not be completely avoided.

In addition to the above changes, instrumental connections were made as follows:
(a) A piezo-electric transducer of AVL make was mounted on the cylinder head through a pressure pick-up hole of 100 mm length and 2 mm dia. The dimension of the connecting hole were calculated according to IS:10105-1982 [15]. A timing wheel was mounted adjacent to the flywheel with a magnetic pulse transducer for measurement of TDC and crankangle marks. Both transducers were led to an AVL Indicating System comprising a Tektronics Storage Oscilloscope and a Polaroid Camera.
(b) An NTC resistor was fixed at the position shown in Fig.4 in the piston. The position was selected to ensure that the measured temperature was always within the NTC resistor range of 150-350°C. The resistor was connected through a non-contact inductive pick-up to the Karl Schmidt peak voltage indicator which was calibrated in terms of piston temperature.
(c) Chromel-Alumel thermocouples were fixed in drilled holes filled with copper powder in the cylinder liner and cylinder head, about 3 mm distant from the inner surfaces.
(d) Chromel-Alumel thermocouples were also fitted at different points in the coolant circuits.

The thermocouple outputs were measured by an SKF digital millivoltmeter.

Other conventional instrumentation such as remotely controlled eddycurrent dynamometer, digital tachometer, AVL needle lift transducer, constant volume bulb with photocell sensors for fuel consumption measurement, Bosch smoke meter, SKF exhaust gas thermometer, etc, were available. A Lawrence and Scott motoring dynamometer was used for motoring tests.

ELECTRICAL ANALOG FOR PISTON - A 3 X half cross-sectional model of the cast iron piston from the crown to the top of the gudgeon pin was built up. The axis was perfectly insulated, and 28 copper plates of 0.5mm thickness were used to define the rest of the boundary. The length of each piece was so determined that the temperature variation along each piece was small. The model was immersed in a rectangular electrolytic tank filled with water acting as electrolyte. The resistivity of the electrolyte was independently measured across a set of two parallel plates. The resistances equivalent to the product of the local heat transfer coefficient α and the local area A were calculated and set on a board of variable resistors. A 400 Hz power supply was used to simulate the temperature in such a manner that

permanent electrolysis and polarisation was avoided. A stiff electrode was used in conjunction with a digital AC voltmeter (50-500 Hz) to probe the voltage in the electrical field. The measured voltages were then converted to temperature values. The theory, techniques and procedure were very similar to those of French [16]. For the present, the Analog is capable of giving a 2-dimensional solution. Modifications to provide an axi-symmetric solution are in progress.

The boundary conditions were assumed as follows:
(a) Piston crown, gas side; The mean effective gas temperature $\overline{T}_g$ and the mean heat transfer coefficient α_m were taken from the cycle simulation programme. It was found that for the coolest and hottest cases considered, α_m varied very little from 0.30 to 0.31 kJ/m^2-s-°K, and $\overline{T}_g$ from 1166 to 1296 °K. An additional assumption was made that the local heat transfer coefficient had a parabolic variation with the cylinder radius:

$$\alpha = \left[1.0 + W\left(0.5 - \frac{r^2}{R^2}\right)\right] \qquad (16)$$

where W = a parabolicity factor which depends on the combustion chamber geometry. Initial trials with the electrolytic tank to match the temperature distribution with the measured values in an aluminium piston showed $W = 1.0$ for the engine under consideration. The same value was used for the cast iron piston for which the combustion chamber geometry was identical.
(b) Piston top land, and upper and back face of top ring groove to liner wall, $\alpha = 0.8$ kJ/m^2-s-°K
(c) Lower face of top ring groove to liner wall, $\alpha = 11.7$ kJ/m^2-s-°K
(d) Lower face of 2nd and 3rd ring grooves to liner wall, $\alpha = 4.1$ kJ/m^2-s-°K
(e) All other surfaces on piston OD and ring grooves, $\alpha = 0.7$ kJ/m^2-s-°K
(f) Piston crown underface to lubricating oil and to gudgeon pin, $\alpha = 0.814$ kJ/m^2-s-°K
The coefficients (b) to (f) are from Munro and Griffith [17] except for (b) which had to be changed from 0.7 to 0.8. The liner wall temperature was an independent variable.

SWIRL TEST RIG - The variation of swirl number and co-efficient of flow against valve lift was measured in a paddle wheel type swirl flow test rig. The dimensions of the rig and the procedures used were the same as Ricardo's [18]. The measured values of swirl number were multiplied by a factor 1.3 to get equivalent impulse meter values [19] from which the non-dimensional input angular momentum N_F was calculated according to the definition:

$$N_F = \dot{J}_R \Big/ \dot{m}_R \, v_{iR} \, R_V \qquad (17)$$

where $\dot{J}_R$ = Rate of input angular momentum in rig

$\dot{m}_R$ = Mass rate of air inlet flow in rig

R_V = Radius of axis of inlet valve from cylinder axis [Fig.1]

v_{iR} = Inlet air velocity in rig

Variation of the factor 1.3 from 0.975 to 1.625 showed no difference to the trends in λ_t, λ, u', and S (Fig. 5). Hence the value of 1.3 was used for further work.

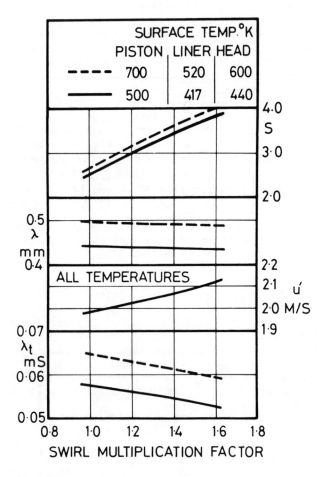

Fig. 5. Effect of varying swirl multiplication factor on flow parameters

PROCEDURE

Engine experiments were conducted by trying to vary the temperature of the liner wall, cylinder head wall, and piston independently by varying the coolant flow to each part. However the temperatures had a certain amount of mutual interaction due to which all three temperatures varied simultaneously. It is proposed in due course to apply statistical techniques to analyse the data which has been obtained. For the

present, another method has been used which is described below under DISCUSSION. During the trials, a record was maintained of the three wall temperatures as well as of the P - Ø diagram, brake horsepower, specific fuel consumption, exhaust gas temperature and exhaust smoke. All tests were conducted with a constant fuel rate of 1.05 kg/h. The base mechanical efficiency of the engine was measured by motoring. The beginning of dynamic injection was measured by the nozzle needle lift transducer.

Using the Electrical Analog, a correlation was obtained between the temperature at the point of measurement shown in Fig.4, and the mean piston crown surface temperature (Fig.6). From this graph, the mean piston crown surface temperature was read off for every measured piston temperature. The mean surface temperatures of the cylinder liner and cylinder head were obtained by a simple heat balance which resulted in an equation of the form:

$$\overline{T_W} = \frac{\propto \overline{T_g} + \frac{k}{t} T_M}{\propto + \frac{k}{t}} \qquad (18)$$

where $\overline{T_W}$ = Mean surface temperature, ^oK

$\overline{T_g}$ = Mean effective gas temperature, ^oK,

T_M = Measured temperature, ^oK

$\propto$ = From Eq.16, kJ/m^2-s-^oK

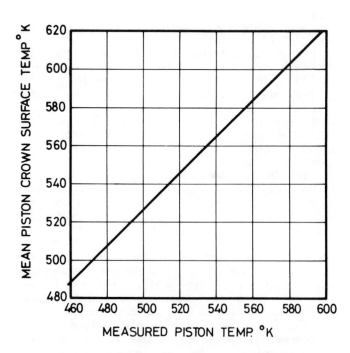

Fig. 6. Correlation between the mean piston crown surface temperature and the temperature at the point of measurement, as measured in Electrical Analog

k   = Thermal conductivity of cast iron
     = 0.05 kJ//m-s-$^{\circ}$K

t   = Distance of point of measurement
     from surface, m

The measured P - $\varnothing$ diagrams were intended to be used to calculate the rate of heat release. The results of this calculation will be reported later.

The K-ϵ model was used to calculate the mixing time λ_t, the mixing length λ, the turbulence velocity u', the squish velocity S_q and the swirl number S. Calculations were done for a range of piston surface temperatures of 500 to 700 $^{\circ}$K, liner surface temperatures of 417 to 520°K and cylinder head bottom suface temperatures of 440 to 600°K. A typical cyclic variation of λ_t and S is illustrated in Fig.7. For purposes of analysis, the values at 11° BTDC representing the approximate beginning of combustion were used.

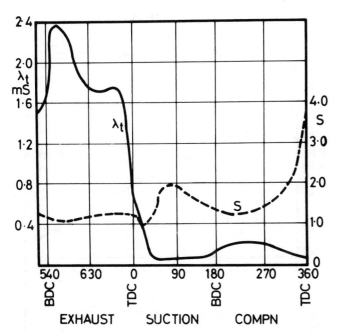

Fig. 7. A typical variation of turbulent mixing time and swirl ratio in the cup against crankangle

One of the theories considered was that the piston crown clearance would decrease due to piston expansion, and that this would cause an improvement of turbulence and swirl characteristics. To test this theory, the K-ϵ model was used to calculate the effect of varying piston crown clearance from 0.5 to 1.0 mm under two conditions, (a) keeping the compression ratio constant at 16.5:1, and (b) allowing the compression ratio to change automatically when the piston crown clearance

was varied. Condition (b) approximates the natural case of piston expansion. The finite element programme was used to estimate the piston expansion at mean piston surface temperatures of 507 and 549 $^{\circ}$K which represent normal and moderately hot pistons respectively. The temperature distribution in the piston required as input was obtained from the Electrical Analog and the peak pressure from the cycle simulation programme.

DISCUSSION

The discussion is divided into two sections dealing with piston crown clearance and wall temperatures. As a rule, both swirl and turbulence should have optimum values with respect to combustion.

PISTON CROWN CLEARANCE - Fig.8 shows the effect of varying piston crown clearance on swirl and squish while Fig.9 shows the effect on λ_t, λ and u'. All calculations have been done for mean surface temperatures of piston 600 $^{\circ}$K, liner 465 $^{\circ}$K and cylinder head 525 $^{\circ}$K. It can be seen that for large changes in piston crown clearance, the influence is quite significant. However the calculations of piston expansion showed that the rim of the cup expands by 0.086 mm at a mean piston crown surface temperature of

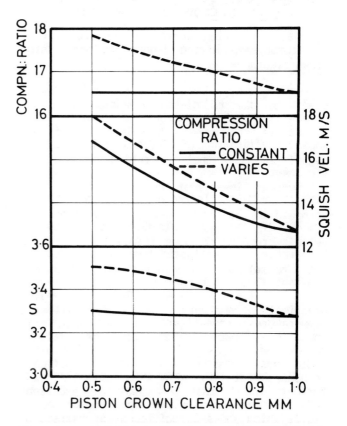

Fig. 8. Variation of swirl ratio and squish velocity against piston crown clearance

507°K and 0.110 mm at 549 °K. For the reduction of the piston crown clearance by 0.11mm, fig. 8 shows that the swirl ratio does not increase at all and the squish velocity increases by 2%. From fig.9, it can also be seen that λ_t decreases by less than 5% and λ, the mixing length, by barely 1%. This analysis suggests that while the trend is favourable, the change in piston crown

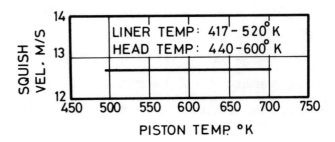

Fig. 10. Variation of squish velocity with piston crown and cylinder head surface temperatures

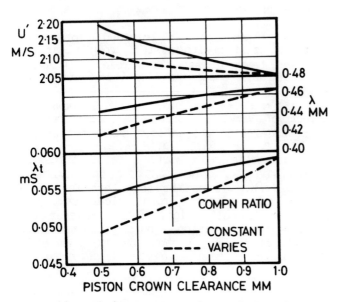

Fig. 9. Variation of turbulence mixing time and turbulence mixing length against piston crown clearance

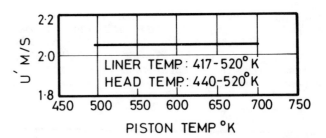

Fig. 11. Variation of turbulence velocity with piston crown and cylinder head surface temperatures

clearance due to piston expansion does not affect the mixing process significantly. The effect may be more with aluminium pistons whose coefficient of thermal expansion is higher although their operating temperatures are lower.

Further investigations were carried out with a constant clearance of 1.0 mm which is standard in the engine under consideration.

EFFECT OF WALL TEMPERATURES – Squish velocity remains constant (Fig.10) for all the temperatures investigated because the density and temperature have been assumed uniform and constant throughout the working space at every instant. The turbulence velocity remains practically constant (Fig.11) for all the temperatures investigated. The result is that λ_t becomes directly proportional to λ (See Eqs.12 and 13). Hence λ_t and S could be considered as the significant responses for this investigation. It was further found that λ_t had an excellent negative correlation with S, as illustrated in Fig.12 which shows a coefficient of regression of 0.98. The plotted points are those of λ_t and S at the measured combinations of temperatures shown in Table 2.

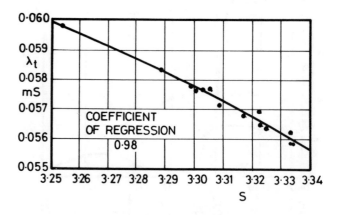

Fig. 12. Correlation of turbulent mixing time and swirl ratio in the cup for various measured testbed conditions

At any constant piston and cylinder head temperatures, the liner temperature had the largest influence on both λ_t and S (Fig.13). The lower the liner temperature, the lower the mixing time, and the higher the swirl. The cylinder head and piston temperatures have smaller effects (Figs.14 and 15 respectively). The higher the surface temperature, the higher the mixing time and smaller the swirl. The piston and cylinder head have approximately equal influence for

Table 2 - Testbed And Computer Results Summary

Sr. No.	Surface Temperature Piston Crown °K	Cyl. Head °K	Liner °K	BSFC g/kWh	Exh. gas temp. °K	Exh. Smoke Bosch No.	Power/ Speed kW/rpm	FMEP bar	IMEP bar	λ_t ms	S -
1	531	455	428	255.4	688	1.6	3.97/1500	1.69	7.82	0.0562	3.333
2	543	461	432	250.0	689	1.9	3.97/1500	1.68	7.80	0.0565	3.323
3	551	475	439	256.8	709	1.2	3.96/1497	1.65	7.79	0.0569	3.322
4	550	474	436	258.0	710	1.6	3.99/1507	1.66	7.80	0.0568	3.317
5	551	491	442	260.8	711	1.6	3.83/1502	1.64	7.54	0.0571	3.309
6	556	496	450	271.0	716	1.8	3.83/1500	1.61	7.53	0.0576	3.300
7	555	496	452	274.5	713	1.8	3.75/1500	1.60	7.40	0.0577	3.299
8	551	463	430	257.5	689	1.9	3.98/1502	1.70	7.82	0.0563	3.324
9	566	475	451	256.8	712	1.7	3.83/1500	1.61	7.52	0.0576	3.302
10	538	460	422	256.8	688	1.6	3.97/1500	1.71	7.56	0.0558	3.333
11	539	464	422	261.0	689	1.6	3.97/1504	1.71	7.57	0.0559	3.332
12	557	467	452	263.0	688	2.0	3.83/1502	1.60	7.24	0.0576	3.305
13	615	510	482	288.0	710	2.1	3.68/1500	1.44	6.87	0.0598	3.254
14	584	475	462	266.0	705	2.2	3.98/1500	1.52	7.37	0.0583	3.288

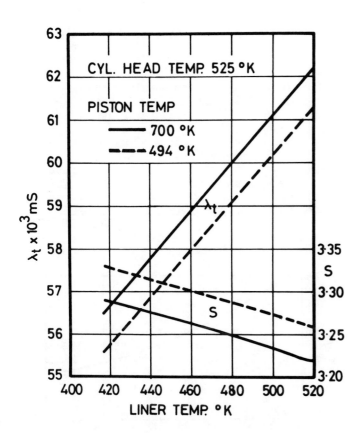

Fig. 13. Variation of cylinder liner temperature vs mixing time and swirl ratio

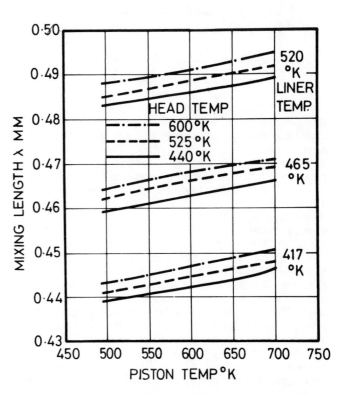

Fig. 14. Variation of mixing length with piston crown and cylinder head surface temperatures

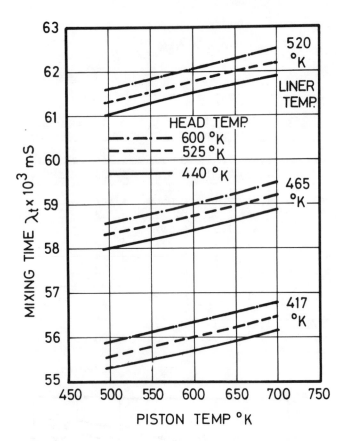

Fig. 15. Variation of mixing time with piston crown and cylinder head surface temperatures

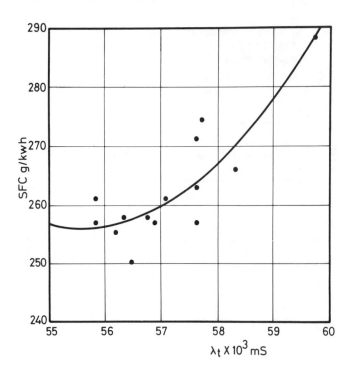

Fig. 16. Variation of SFC with mixing time

the same change of temperature. These results show that swirl is increased and mixing time reduced by lowering the temperatures of the piston, cylinder head and especially the liner. The influence is about the same for large changes in piston and cylinder head temperatures and for small changes in liner temperatures.

In the testbed experiments all three wall surface temperatures changed simultaneously for each experiment. For each set of values of the three temperatures, the corresponding values of λ_t and S were calculated. The base FMEP of the engine was corrected for each run by assuming that only the influence of the liner temperatures on the lubricating oil viscosity need be considered to estimate the change in mechanical efficiency between run and run. Combined with the measured BMEP, the IMEP was estimated. Table 2 shows a summary of the testbed and computer results, the latter being shown in the last two columns.

Fig. 16 shows the plot of λ_t vs BSFC and fig.17 shows S vs BSFC. The lines drawn are quadratic approximations with a coefficient of regression of 0.75 and 0.77 respectively. The results show that low wall temperatures promote mixing. A similar

analysis, for exhaust smoke density failed to show any significant correlation. For IMEP, the coefficient of regression was about 0.76. The plot of measured values of IMEP vs λ_t is shown in Fig.18 from which one may infer that with low temperatures, there is a tendency towards improved IMEP, the optimum λ_t being about 0.056.

Figs.16 and 17 show that BSFC also reaches a minimum when λ_t is about 0.056,when S is about 3.32. What is notable is the very high sensitivity of SFC and MEP to a variation in λ_t and S. A change of barely 1% in either λ_t or S is enough to upset the engine performance. It can be inferred that the flow parameters considered in this paper are not the primary cause for the improved

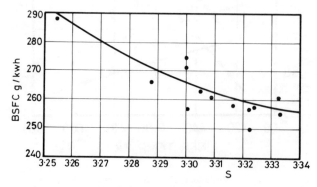

Fig. 17. Variation of SFC with swirl ratio in the cup

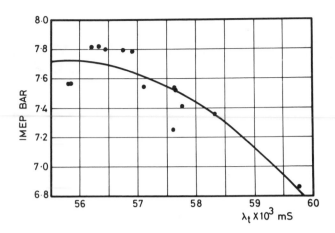

Fig. 18. Plot of IMEP against mixing time for different runs

performance of differentially cooled engines. However it is known that λ_t and S should have optimum values in a given engine. The present investigations show that the optimum λ_t may be around 0.056 and optimum S around 3.32 for the engine under consideration..

Referring back to Figs. 13 and 15, one can see that to maintain the optimum conditions, the liner wall temperature should be maintained around 430°K or less while the piston temperature and cylinder head temperature bear a reciprocal relationship to each other. A higher piston temperature of about 660°K should be combined with a low cylinder head temperature of 440°K while a low piston temperature of 500°K should be combined with a high cylinder head temperature of 600°K. The values indicated refer to the engine investigated although similar trends may appear in other naturally aspirated engines.

The conditions regarding surface temperatures shown here are necessary but not sufficient as shown by the low coefficient of regression in the points plotted in Figs. 16,17 and 18. This suggests that there are other factors of more importance than flow factors playing their part in the phenomenon under study. Such factors may include, for example, evaporation, wall impingement, spray formation, fuel properties, etc. These factors are being studied, initially with order-of-magnitude calculations. However a high piston temperature did show a lower SFC earlier [1, 2]. The present investigation shows that this should preferably be combined with low cylinder head and liner temperatures. Such a system is possible if an insulated piston is used in a conventionally water-cooled engine. The low liner temperature is also advantageous to efficient piston ring operation and cylinder liner durability.

CONCLUSIONS

From this investigation, the following conclusions may be drawn for the naturally aspirated engine under study.
(1) For optimum mixing conditions, the liner wall temperature should be around 430°K or less.
(2) The cylinder head and piston temperatures bear a reciprocal relationship to each other. A high piston temperature should be associated with a low cylinder head temperature and vice versa in order to promote efficient mixing.
(3) These relationships are necessary for efficient combustion but not sufficient. Other factors such as fuel evaporation, wall impingment, spray formation, fuel properties, etc, need further study.
(4) A high piston temprature has been shown earlier to be favourable. The present investigation shows that this should preferably be combined with low cylinder head and liner temperatures.

ACKNOWLEDGEMENTS

The work reported in this paper is the result of a group effort. However the contributions of some individuals have been outstanding and are gratefully acknowledged. Dr. P.A. Lakshminarayanan developed the swirl and turbulence programme and helped in the analysis. With the assistance of Mr. D.S. Gunjegaonkar, he constructed the Electrical Analog and conducted the experiments on it. Dr. S.S. Patil handled the finite element programme. Mr. U.P. Nagpurkar built the engine and conducted engine trials on the dynamometer testbed. Mr. S.N. Damle and Mr. L.V. Rajaram assisted in the statistical analysis.

The authors also thank the management of Kirloskar Oil Engines Limited for permitting them to publish the contents of this paper.

REFERENCES

[1] Kirloskar, C.S. et al: The AV1 Series III Diesel - A Differentially Cooled Semi - Adiabatic Engine Below 10 kW, SAE Paper No. 790844, USA, 1979.
[2] Kirloskar, C.S. et al: Differential cooling - A New System for Higher Economy in Small Diesels, CIMAC Paper No. D 4.2, 15th International Congress on Combustion Engines, Paris, 1983.
[3] Watson, N et al: A Combustion Correlation for Diesel Engine Simulation, SAE Paper No. 800029, USA, 1980.
[4] Wolfer, H: Der Zunderzug im Dieselmotor, VDI-Forschungsheft No.392, 1938, pp 15-24.
[5] Woschni, G: Die Berechnung der Wanderluste und der Thermischer Belastung der Bauteile von Dieselmotoren, MTZ, Vol.31, No.12. 1970.

[6] Krieger, R.B, and G.L. Borman : The Computation of Apparent Heat Release for Internal Combustion Engines, ASME Paper No. WA/OGP-4, USA, 1966.

[7] Millington, B.W, and E.R. Hartles: Frictional Losses in Diesel Engines, SAE Paper No,680590, USA, 1968.

[8] Zurmühl, R: Praktische Mathematik für Ingenieure und Physiker, Springer Verlag, Berlin, 1965.

[9] Borgnakke, C et al: Prediction of In-Cylinder Swirl Velocity and Turbulence Intensity for an Open Chamber Cup in Piston Engine, SAE Paper No, 810224, 1981.

[10] Launder, B.E. and D.B. Spalding: Lectures in Mathematical Models of Turbulence, Academic Press, 1972.

[11] Schlichting, H: Boundary Layer Theory, Mc Graw Hill, New York, 1980.

[12] Davis, G.C, and J.C. Kent: Comparison of Model Calculations and Experimental Measurements of the Bulk Cylinder Flow Processes in a Motored PROCCO Engine, SAE Paper No. 790290, USA, 1979.

[13] Tennekes, H and J.L. Lumley: A First Course in Turbulence, The M.I.T. Press, London, 1972.

[14] Wilson, E.L: Computer Programme for the Analysis of Axi-Symmetric Solids, University of California, Berkeley, Feb 1967.

[15] IS:10105-1982: Specification for Fittings for Cylinder Pressure Indicators for Internal Combustion Engines, Indian Standards Institution, New Delhi, July 1982

[16] French, C.C.J, and M.L. Monaghan: Loading of Highly Rated Two-Cycle Marine Diesel Engines, CIMAC Paper No.A8, 6th International Congress on Combustion Engines, 1965

[17] Munro, R, and W.J. Griffiths: Diesel Piston Design and Performance Predictions, 11th International Congress on Combustion Engines, Barcelona, 1975, pp. 369-400.

[18] Anon: Port Blowing Rig - Working Method, Report No. DP.79/300 U, Ricardo Consulting Engineers, UK (undated).

[19] Partington, G.D: Analysis of Steady Flow Tests on Inlet and Exhaust Ports, Report No.DP 80/1123, Ricardo Consulting Engineers, U.K. July 1980.

840428

Cummins/TACOM Advanced Adiabatic Engine

R. Kamo
Cummins Engine Company, Inc.
W. Bryzik
U.S. Army Tank-Automotive Command

ABSTRACT

Cummins Engine Company, Inc. and the U.S. Army have been jointly developing an adiabatic turbocompound engine during the last nine years. Although progress in the early years was slow, recent developments in the field of advanced ceramics have made it possible to make steady progress. It is now possible to reconsider the temperature limitation imposed on current heat engines and its subsequent influence on higher engine efficiency when using an exhaust energy utilization system.

This paper presents an adiabatic turbocompound diesel engine concept in which high performance ceramics are used in its design. The adiabatic turbocompound engine will enable higher operating temperatures, reduced heat loss, and higher exhaust energy recovery, resulting in higher thermal engine efficiency. This paper indicates that the careful selection of ceramics in engine design is essential. Adiabatic engine material requirements are defined and the possible ceramic materials which will satisfy these requirements are identified. Examples in design considerations of engine components are illustrated. In addition to these important points, the use of ceramic coatings is described in the design of engine components. The first generation adiabatic engine with ceramic coatings is described. The advanced adiabatic engine with minimum friction features utilizing ceramics is also presented. The advanced ceramic turbocharger turbine rotor as well as the oilless ceramic bearing design is described. Finally, the current status of the advanced adiabatic engine program culminating in the AA750 V-8 adiabatic engine is presented.

THE CONTEMPORARY DIESEL ENGINE has served the world well for many decades. However, the emissions regulations of the late 60's and the energy crisis of the early 70's have prompted Cummins Engine Company, Inc. of Columbus, Indiana, and the U.S. Army Tank-Automotive Command of Warren, Michigan, to undertake a joint program to take a giant step toward improving the energy and material conservation efforts of the future vehicular power plants. These efforts were not to compromise with engine emissions characteristics. The effort led to the adiabatic turbocompound and the advanced minimum friction engine development programs.

The feasibility of the adiabatic turbocompound engine was demonstrated in February, 1980, and the efforts led to a first generation adiabatic engine installed in a five-ton U.S. Army truck. Subsequently, in February, 1981, the program undertook the prototype development of a Cummins 325 HP Vee-eight commercial truck engine into a 700 HP adiabatic turbocompound engine. The prototype engines are scheduled for completion in August, 1984. These prototype engines will be insulated with ceramics and will entail no cooling system whatsoever.

The prospects of using ceramics for more advanced engine concepts appeared attractive. The minimum friction engine (MFE) was conceived. The MFE is expected to operate without any lubrication and reduce the total engine friction by 50%. The development status of this engine concept is also delved upon.

Whatever the engine cycle or the engine configurations may be for the future, it is quite apparent that higher cycle pressures and temperatures will be the way. This paper will cover the role of the high technology fine ceramics in future engine concepts. It will attempt to point out the problem areas and the technical approach to overcome them. Engine operating environments and the property requirements of ceramics based on 10 years development experience are presented. Some of the current promising ceramics applied to selected engine components are also covered.

THE ADIABATIC TURBOCOMPOUND DIESEL

A simplified schematic of the Adiabatic Engine is shown in Figure 1. Following the engine flow path, air enters the turbocharger (is compressed) and then enters the insulated, high-temperature combustion chamber of the piston unit. Insulated combustion chamber components include those previously noted. Combustion occurs and useful energy is extracted from the piston unit. The high-temperature, high pressure exhaust gas is then expanded through two turbine wheels to extract as much of the remaining energy as possible. One wheel is used to drive the compressor, and the second is connected by gears (turbocompounding system) to the engine crankshaft to further increase the useful power output of the engine. It should be noted that the adiabatic turbocompound diesel will require no engine cooling system.

OPERATING ENVIRONMENT (1)

The operating temperature within the insulated combustion chamber of the adiabatic engine depends greatly on the load, air-fuel ratio, intake air temperature, injection timing, etc. Figure 2 shows a least square curve fit of the thermocouple probe temperature within the combustion chamber. The thermocouple probe temperature shown in Figure 2 is thought to provide a reasonable approximation of time-mean gas temperature of an adiabatic engine (12). Thus, the measured temperature was used in all piston, liner, and head analyses. For design analysis purposes, the peak cylinder pressure is 2000 psi. A typical combustion diagram as a function of crank angle is shown in Figure 3. The 5.5 inch bore direct injection engine from which the pressure diagram was taken will have a peak load of 25 tons on the piston top. Figure 4 shows the expected temperature distribution within the combustion chamber of an adiabatic engine with a zirconia cylinder liner backed with a cast iron sleeve.

The engine under investigation had the following operating specifications:

Bore x Stroke	5-1/2 x 6 inches (140 x 152 mm)
Engine Speed	1900 rpm
Engine Configuration	In-line Six cylinder
Engine Cycle	4-Cycle
Combustion Chamber	Direct Injection Quiescent Chamber
Peak Cylinder Pressure	1800 psi
Lubrication	Oil Lubricated
Brake Mean Effective Pressure	195 psi
Air-Fuel	27:1
Intake Air Temperature	140°F
Overspeed Capability	25% Burst Margin

CERAMICS IN ENGINE

The operating environments of an adiabatic diesel engine were shown in the previous section. The pressure, temperature, reliability, and durability requirements are quite stringent. With higher BMEP (brake mean effective pressure) output trends of diesel engines (Figure 5), ceramic components must continue to withstand these higher pressures and temperatures if the adiabatic engine concept is to be adopted.

An all ceramic engine is still in the distant future, and the approach considered in the adiabatic engine program at Cummins is the composite approach. The composite approach is the use of suitable ceramic materials in the hot combustion and exhaust zone together with cast iron or aluminum engine parts. It is quite obvious that with the

composite design, the ceramic material coefficient of thermal expansion should be similar to the metal. Otherwise, a compliant layer must be used to overcome the mismatch in thermal expansions. Therefore, a ceramic material with high coefficient of expansion corresponding to iron or aluminum is highly desirable.

DESIRED CERAMIC PROPERTIES (13)

First, an adiabatic engine is one in which no heat is added or subtracted in the thermodynamic cycle. A perfect insulating material of ceramic is highly desirable. However, such materials do not exist and, therefore, the best available insulating material within reason must be considered. The early high performance ceramics, developed for advanced gas turbines, were silicon carbide and silicon nitride; but they fell short in the area of insulation properties.

Second, the ceramic material being considered for the adiabatic engine must possess high temperature strength. The high brake mean effective pressure trend of the diesel engine was emphasized. Peak cylinder pressures over 2,000 psi and surface temperatures approaching 1000°C (1800°F) are common. Any candidate material must be able to withstand these pressures and temperatures. Most important, however, is the thermal stress which develops in a design which could be much greater than the mechanical loading. An example is shown in Figure 6 for the case of an insulated piston design. In this piston design, the maximum tensile stress at the bottom of the piston which was induced as a result of the thermal gradient was about 65 ksi, while the stress at this location was only increased by 20 ksi due to cylinder pressure.

In the case of the cylinder liner, additional functions which cannot be overlooked for optimized engine operation are friction and wear properties. Corrosion and erosion also cannot be overlooked. Materials with low density are also desired because of inertia and weight factors.

Aging properties of materials become highly critical. Partially stabilized zirconia (PSZ) with MgO stabilizer is prone to change phase at elevated temperatures. Transformation changes can be avoided by use of yttrium oxide stabilizer.

In general, the use of ceramics in composite designs for the metallic diesel engine, the following ceramic properties are desired:
. Good Heat Insulation
. High Expansion Coefficient
. High Temperature Strength
. Low Wear/Corrosion/Erosion
. Low Friction Characteristics
. High Hertz Stress/Fatigue Durability
. Low Cost/Weight
. Involve No Strategic Materials
. Close Tolerances and Fine Finishes
. Good Dimensional Stability
. Low Density (inertia)
. Limited Plasticity (creep)
. Good Thermal Shock Resistance
. High Fracture Toughness

CUMMINS NH TURBOCOMPOUND DIESEL ENGINE

A hybrid diesel–turbine system in which piston power is supplemented by turbine power recovered from the exhaust gas.

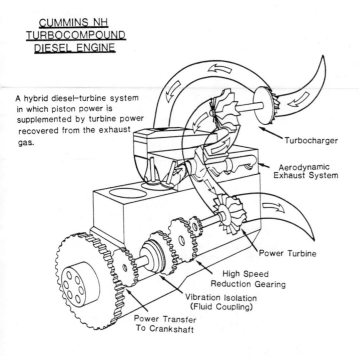

FIGURE 1 - SKETCH OF THE CUMMINS TURBOCOMPOUND DIESEL ENGINE

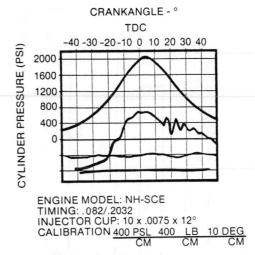

ENGINE MODEL: NH-SCE
TIMING: .082/.2032
INJECTOR CUP: 10 x .0075 x 12°
CALIBRATION $\frac{400 \text{ PSL}}{\text{CM}} \quad \frac{400 \quad \text{LB}}{\text{CM}} \quad \frac{10 \text{ DEG}}{\text{CM}}$

1900 RPM AT FULL LOAD

FIGURE 3.

Combustion Diagram of an Uncooled 5 1/2″ x 6″ (Bore x Stroke) Diesel Engine Operating at Full Load and Speed

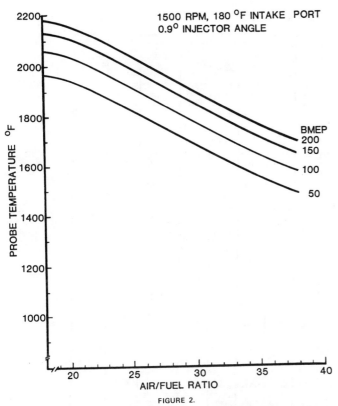

1500 RPM, 180 °F INTAKE PORT
0.9° INJECTOR ANGLE

FIGURE 2.

Least–square fit of probe temperature

FIGURE 4 - **Predicted Piston and Zirconia/ Cast Iron Liner Temperature**

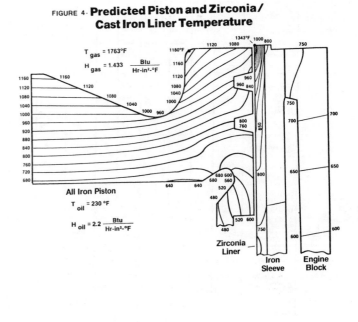

$T_{gas} = 1763°F$

$H_{gas} = 1.433 \frac{Btu}{Hr\text{-}in^2\text{-}°F}$

All Iron Piston

$T_{oil} = 230 °F$

$H_{oil} = 2.2 \frac{Btu}{Hr\text{-}in^2\text{-}°F}$

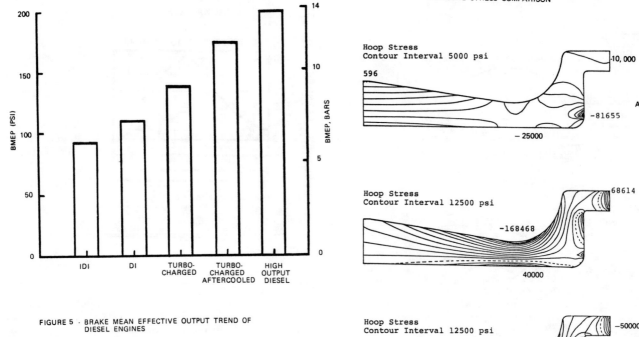

FIGURE 6 - PISTON CAP ASSEMBLY, THERMAL, AND CYLINDER PRESSURE STRESS COMPARISON

FIGURE 5 - BRAKE MEAN EFFECTIVE OUTPUT TREND OF DIESEL ENGINES

TABLE 2.
PROPERTIES OF CERAMICS AND METALS

Material / Property	Alumina	Cordierite Low Density	Cordierite High Density	Mullite	Zirconia Stabilized	Zirconia PSZ	Electrical Porcelain	Silicon Nitride Pressureless Sintered	Silicon Nitride Hot Pressed	Silicon Nitride Reaction Bonded	Sialon Reaction Sintered	Silicon Carbide Pressureless Sintered	Silicon Carbide Hot Pressed	Silicon Carbide Self Bonded	SUS 304 (AISI-304)	Nimonic 105	FC 25 (Gray Cast Iron SAE G3500)	AC 4C (SAE 356.0)
Bulk Density	3.98	1.61	2.50	3.26	5.4	5.91	2.45	3.08	3.18	2.75	3.00	3.10	3.21	3.10	7.93	7.99	7.0	2.69
Open Porosity (%)	0	37	0	0	0	0	0	0	0	6	0	0	0	0	0	0	0	0
Flexural Strength RT / 1,000°C (MPa) (4-Point)	440 / 340	13 / 13	70 / 70	172 / 98	186 / –	1020 / 400	157 / –	650 / 470	845 / 680	296 / 300	355 / 355	500 / 475	930 / 820	530 / 530	755 (Tensile)	972 (Tensile)	240 (Tensile)	240 (Tensile)
Young's Modulus (GPa)	360	16	118	144	160	205	94	230	310	160	231	303	440	410	193	175-224	80-110	72
K_{IC} (MN/m$^{3/2}$)	4.5	–	--	–	1.1	8.4	1.6	5.3	5.6	3.6	2.2	2.4	4.4	4.6	(80-100)	(80-100)	(80-100)	–
Creep Resistance (MPa)	100 (1,000hr / 1,200°C)	–	–	60 (1,000hr / 1,200°C)	–	–	–	280 (1,000hr / 1,000°C)	–	250 (1,000hr / 1,000°C)	–	350 (1,000hr / 1,000°C)	–	–	70 (1,000hr / 800°C)	100 (100,000hr / 800°C)	–	165 (1,000hr / 150°C)
Linear Thermal Expansion Coefficient (×10^{-6}/°C)	8.1	1.2	2.0	5.13	10.9	10.5	6.5-7.4	3.3	3.28	3.0	3.0	4.3	4.8	4.3	17.3	12.5	10-11	23.5
Thermal Conductivity (cal/cm·sec·C)	0.07	0.003	0.010	0.016	0.005	0.007	0.004	0.070	0.048	0.048	–	0.14	0.19	0.16	0.039	0.028	0.12	0.38
Heat Capacity (cal/g·C)	0.2	0.12	0.17	0.15	0.18	0.12	0.28	0.19	0.17	0.19	–	0.20	0.20	0.20	0.12	0.11	0.13	0.23
Thermal Shock Resistance Parameter R = St/E·α	150	680	300	230	110	350	240	860	830	620	510	380	440	300	–	–	–	–
Oxidation Resistance (mg/cm^2)	–	–	–	–	–	–	–	0.1 (100hr / 1,000°C)	<0.1 (100hr / 1,000°C)	2 (100hr / 1,000°C)	–	<0.1 (100hr / 1,000°C)	1.5 (16hr / 1,600°C)	–	–	–	–	–
Melting Point (C)	2,030	1,440	1,450	1,810	2,600	2,600	1,100	1,900 (decomp.)	1,900 (decomp.)	1,900 (decomp.)	–	2,700 (decomp.)	2,700 (decomp.)	2,700 (decomp.)	1,400	–	1,150	555
Max. Use Temperature (C)	1,800	1,200	1,200	1,600	2,300	1,500	500	1,400	1,500	1,500	–	1,650	1,650	1,650	–	–	–	–

In quantifying some of the ceramic properties for use in the adiabatic diesel, Table 1 indicates the desired values based on nine years of background and design experience with ceramics.

Table 1 - Typical Desired Material Properties for an Adiabatic-Type Diesel Engine

Temperature Limit, °C	>1800
Fracture Toughness, $MN/M^{3/2}$	> 8.0
Flexural Strength, MPa	> 800
Thermal Conductivity, Cal/CM-SEC-°C	<0.01
Thermal Shock Resistance, T °C	> 500
Coefficient of Expansion x 10^{-6}/°C	> 10
Weibull Modulus	> 18
Time, Exposure, Hours	>1000

CANDIDATE CERAMIC MATERIALS (13)

There are many new materials now available commercially or in development in laboratories. However, none will meet the desired specifications of the adiabatic engine shown in Table 1. Table 2, provided by NGK Insulators, summarizes the important properties of most common ceramics and some metals used predominantly in the automotive industry.

From the materials table, Si_3N_4 and SiC do not meet the desired insulation and expansion coefficient properties. Alumina is generally low on flexural strength, toughness, thermal shock resistance, and insulative properties. Mullite is good on insulative properties but has inadequate strength. The same is true for cordierites and fused silica.

The material possessing the nearest compromise to the requirements shown in Table 1 is the yttria stabilized partially stabilized zirconia.

From the above, it is noted that the partially stabilized zirconia possesses very good insulation property and compatible coefficient of expansion to match the cast iron engine components. Most current component designs for the adiabatic engine are composed of shrink interference fitted PSZ materials with cast iron. Major components include the piston crown, cylinder head hot plate, valve seat inserts, and cylinder liner.

The MgO stabilized partially stabilized zirconia has exceptionally good fracture toughness and thermal shock resistance. However, aging properties at elevated temperatures leave some things to be desired. Figure 7 shows the phase change in partially stabilized zirconia. The MgO stabilized PSZ tends to revert back to the original monoclinic phase from the tetragonal. Other disadvantages of PSZ would be its high density and cost -- especially when yttria stabilized PSZ is considered.

Although the yttria stabilized PSZ has performed satisfactorily, no long term durability aging properties have been measured. The long term properties that should be determined in any of these promising materials are:
. Phase Change
. High Temperature Creep
. Oxidation and Wear
. Selective Corrosion/Destabilization
. Permeability to Low Quality Fuel Condensate
. Reaction with Metal Components

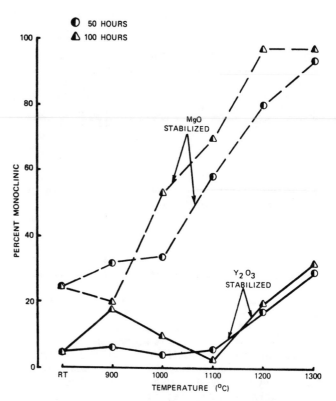

Figure 7 - PHASE CHANGE IN PARTIALLY STABILIZED ZIRCONIA (Ref. 16)

. Stress Rupture and Dynamic Stress Rupture
. Thermal Fatigue
. Corrosions/Deposits

In the future, as ceramic technology advances, the newer materials which could spearhead the material list for the waterless adiabatic diesel have been prioritized by Oakridge National Laboratory and listed in Table 3.

Table 3 - Tentative Priority Listing of Candidate Heat Engine Ceramics by Engine Type

I. Bulk Ceramics
    1. Partially Stabilized ZrO_2 (PSZ)
    2. PSZ + metal or oxide dispersoid
    3. Al_2O_3 + HfO_2 dispersoid
    3. Pressureless sintered Si_3N_4
       (PS Si_3N_4)
    4. Al_2O_3 + ZrO_2 dispersoid
    4. Al_2O_3 + metal dispersoid
    5. Mullite
    5. Mullite + metal or oxide dispersoid
    5. Low thermal expansion ceramics
    6. Si_3N_4/Si_3N_4 composites
       (Fiber/Matrix)
    7. SiC/Si_3N_4 composites
    8. SiC/SiC composites
II. Ceramic Coatings
    1. ZrO_2 base and HfO_2 base
    2. Non ZrO_2 base
    3. Boride
    3. Carbide
    3. Nitride

CERAMIC COATINGS

An alternative to the monolithic ceramic used in composite adiabatic engine components as discussed above is the ceramic coating. There are many new ceramic coating techniques which are quite attractive alternatives to the monolithic ceramic approach.

The most popular ceramic coatings is the plasma spray of zirconia onto a metal substrate with suitable bond coating. Zirconia with yttria additive, i.e. ZrO_2-.08 Y$_2$O$_8$ as a thermal barrier coating and Ni-16.8, Cr-5.8, Al-11.8 Y_2O_3 bond coatings offer an attractive thermal barrier coating system for an adiabatic engine. The coating of aluminum pistons, for example, with ZrO_2 is very difficult and usually ends in failure due to the large thermal expansion coefficient mismatch. Failure by delamination was shown by Miller and Lowell (9) to precede surface cracking or spalling.

One of the difficulties with plasma spray is the inability to coat a thick layer in order to establish a large thermal barrier. Figure 8 shows the effect of ZrO_2 coating thickness on heat loss through the piston and the fuel economy (BSFC) of the engine shown in the section of "Operating Environment". To achieve meaningful insulation results which can reflect an improvement in engine efficiency, a 5 mm (0.200 inch) coating is needed. So far, plasma spray coating technology is in the 0.25 to 0.5 mm (0.010 to 0.020 inch) thickness range. Beyond these values, spalling occurs due to thermal stress. A thermal-stress analysis for a coated iron piston is summarized in Figure 9 (residual coating stresses are not included). Improvements in plasma spray coating have been shown, by increasing the power level and arc gas composition used during plasma spraying, on the life of two-layer thermal barrier systems.

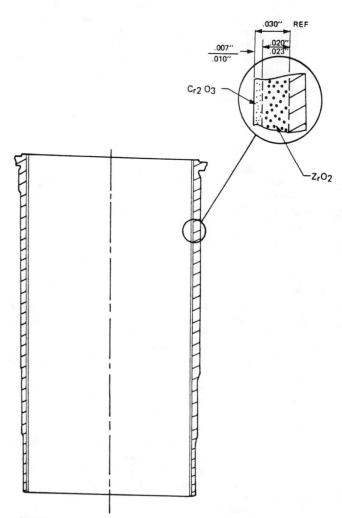

FIGURE 9 – Predicted Thermal and Stress Summary for Plasma Sprayed Zirconia on Iron Pistons

Coating Thickness (in.)	Piston Temperatures (°F)				Heat Loss (Btu/min)	Maximum Coating Stress (ksi)			
	Coating A	Piston B	Interface C	Ring Groove D		Hoop*	Maximum Principle	Bond Stress Shear	Normal
0	1180	640	1160	960	360				
.050	1509	522	702	800	290	38	39		3
.100	1660	443	553	720	220	44	44	4	16
.180	1720	381	421	680	160	60	60	12	18

Analysis Conditions

195 BMEP
T_g = 1763°F
H_g = 1.433 Btu/hr-in^2-°F
T_{oil} = 230°F
H_{oil} = 2.2 Btu/hr-in^2-°F

cylinder liner and its three-layer coated system is shown in Figure 10. The hard wear surface was initially applied by plasma spray but subsequently coated with a post densification process developed by Kaman Sciences Corporation.

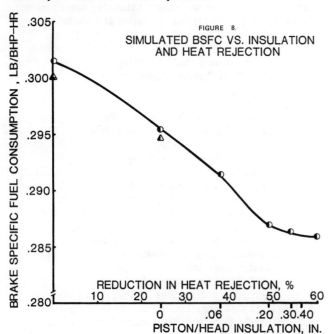

FIGURE 8.
SIMULATED BSFC VS. INSULATION AND HEAT REJECTION

A zirconia coated cylinder liner by itself quickly wears away. A proprietary three-layer coating system was developed by coating a thin 0.125 mm (.005 inch) thick chrome oxide or aluminum titanate wear surface over the zirconia thermal barrier coating. The

FIGURE 10 - COMPOSITE ZrO_2 - Cr_2O_3 COATED CAST IRON LINER

The Kaman Sciences process densifies porous ceramics by depositing chromium oxide, which is in a liquid solution, into the pores and surrounding grains of the ceramic material's structure. The most common means of application is to immerse the component, with the ceramic thermal barrier coating on it, in the solution. Where immersion is impractical, the solution can be sprayed or painted on. The impregnated ceramic is fired to ~550°C (1000°F). Moisture is removed and a chemical reaction takes place, bonding the chromium and zinc oxide to themselves and to the oxide components of the ceramic. A chrome oxide densified silica-chromia-alumina coating can be applied to the base metal when a thermal barrier coating is not needed. It provides excellent wear and chemical resistance. In epoxy pull tests, the bond consistently exceeds 69 MPa, the test limit.

The physical, thermal, and chemical properties of the Cr_2O_3 coating are remarkable. Some of the important properties are shown in Table 4.

Table 4 - Properties of Cr_2O_3 Densified Coating
A - Physical Properties

Thickness	: .002-.003 in. (40-50 microns)
Porosity	: no surface porosity
Hardness	: 1800-2000 Vickers (50 gr LD)
Bond Strength	: 11-12,000 psi (800 kg/cm^2) (test limit of epoxy)
Compressive Strength	: 1.176 (10^6) psi (80 h Bars)
Modulus of Elasticity	: 367.5 (10^6) psi (25 10^3 h Bars)
Bending Strength (flexural)	: 0.294 (10^6) psi (30 h Bars)
Electro Chemical Potential	: 800 mili volts
Electrical Resistance	: 10^5 to 10^6 OHM/M

B - Thermal Properties

Coefficient of Expansion	: 7.5-22 10^{-6}/°C
Thermal Shock Resistance in H$_2$O	: 1500°F (800°C)
Normal Average Use Temperature	: 1300°F (700°C)

In most cases, the coating will withstand as much heat as the substrate that it is on.

C - Chemical Properties

The coating will pass immersion tests without attack in the following solutions:
- sea water
- all chemical base solutions
- most acids
- all principal solvents

The two acids, HYDROCHLORIC and HYDROFLOURIC, have been found to attack the REVETOX coating after 72 hours in high percentage concentrations. However, the coating is successfully used in solutions up to 15%.

The insulating properties of coatings are quite good when compared to its monolithic counterpart because of the high degree of porosity in most coatings. The thermal conductivity of common monolithic insulating ceramics and in the plasma sprayed form as well as the above described K'RAMIC form is shown below. Other common materials such as iron are shown for relative comparison.

THERMAL CONDUCTIVITY
(Btu/in/hr-ft^{2}°F)

Material	Room Temp.	1470°F
Iron	270	259
MgO Stabilized PSZ	22.7	17.01 (800°F)
Y$_2$O$_3$ Stabilized PSZ	16.6	15.4
Plasma Sprayed ZrO$_2$	5.91/9.01	5.67/6.53
Plasma Sprayed ZrO$_2$ Coated with Cr$_2$O$_3$	9.80	9.69

DESIGN CONSIDERATIONS

Three builds of basic adiabatic insulated engine approaches (metal, HPSN, and LAS) were initially designed and constructed. The metallic and the high performance ceramics (HPSN) required 20 layers of roughened 0.25 mm (0.010 inch) thick stainless steel shims in order to insulate while the LAS material was inherently an insulator. The conclusions from these three builds were that glass ceramic materials (LAS) are inadequate in strength in spite of their insulation value. The hot press silicon nitride presented design compatibility problems with it low coefficient of expansion. The metallic build provided low adiabaticity value and will be subject to high temperature fatigue failure at 195 BMEP (psi) loads due to high temperatures, high expansion, and high conductivity.

In June, 1979, at CIMTECH IV in St. Vincent, Italy, (4), we described the high performance ceramic adiabatic engine shown in Figure 11. The engine was insulated by means of insulating shims behind the hot plate and piston cap. A deck spacer sandwiched between two layers of insulating shim packs was used to form liner insulation. Figure 12 shows the most recent design of the same engine by using PSZ material. Considerable simplification can be noted in this engine design.

In developing engine components, finite element technique was extensively used. It will not be possible to cover all the engine components here, but the approach in designing adiabatic components is similar. The case of the interference fitted piston cap was presented in Paris, France, in December, 1982, (11). In this paper, analysis of selected piston cap designs will be presented.

Figure 4 showed the predicted piston and zirconia/cast iron cylinder liner temperatures using thermal boundary conditions as obtained from diesel cycle simulation and the single cylinder test engine described under "Operating Environment". Figure 13 shows the indicator diagram of the cylinder pressure as a function of piston stroke. The great swings in pressures and temperatures indicate the complexity of the problem.

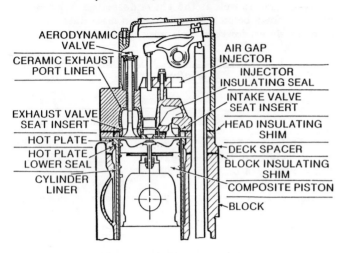

FIGURE 11 - **Cross section of Cummins basic adiabatic (insulated) diesel engine**

AERODYNAMIC VALVE
CERAMIC EXHAUST PORT LINER
EXHAUST VALVE SEAT INSERT
HOT PLATE
HOT PLATE LOWER SEAL
CYLINDER LINER

AIR GAP INJECTOR
INJECTOR INSULATING SEAL
INTAKE VALVE SEAT INSERT
HEAD INSULATING SHIM
DECK SPACER
BLOCK INSULATING SHIM
COMPOSITE PISTON
BLOCK

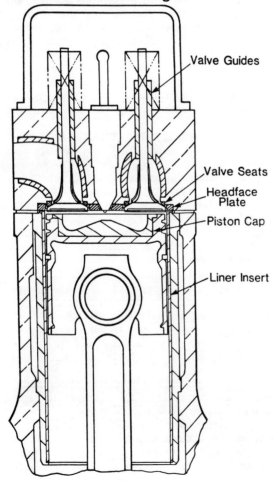

FIGURE 12 - **Zirconia Insulated Engine**

Valve Guides
Valve Seats
Headface Plate
Piston Cap
Liner Insert

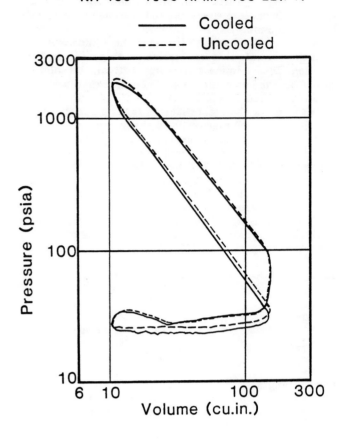

NH 450   1900 RPM/1100 LB.FT.

——— Cooled
- - - - - Uncooled

Figure 13.

Plot Of P-V Diagram

Attachment of a ceramic piston cap for insulating purposes can be achieved by a number of techniques which might include bolting, interference fitting, brazing, or casting into metal. Bolted assemblies were used in the earlier design described in (4). This design was composed of a large number of parts and control of bolt preload was critical. A number of complex variables influenced fatigue of the bolt; and due to low reliability, this approach was eventually dropped. Casting of monolithic ceramics into metal to solve attachment requirements has not yet been fully developed and is primarily controlled by material thermal shock properties. Brazing of ceramics to metals is currently being developed, but not ready for engine testing. The press-fit assembly is the simplest attachment technique and has been utilized to date. The most significant disadvantage of the press-fit design is obtaining an interference fit on the entire cap circumference and achieving insulation of the entire piston top surface. As shown in Figure 14, the primary advantage of full top insulation is in reducing metal temperatures in the ring groove area. The ring groove temperature for full surface insulation is 360°C (660°F) as opposed to partial surface insulation ring groove temperature of 430°C (800°F).

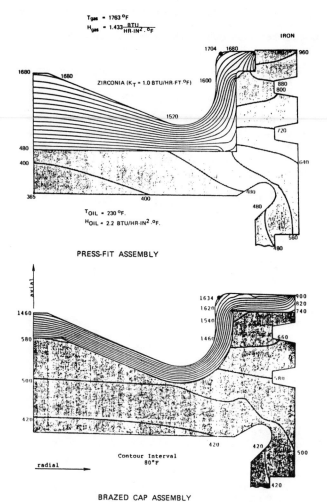

T_gas = 1763 °F
H_gas = 1.433 $\frac{BTU}{HR \cdot IN^2 \cdot °F}$

IRON

ZIRCONIA (K_T = 1.0 BTU/HR·FT·°F)

T_{OIL} = 230 °F.
H_{OIL} = 2.2 BTU/HR·IN²·°F.

PRESS-FIT ASSEMBLY

axial

radial

Contour Interval
80 °F

BRAZED CAP ASSEMBLY

Figure 14 - Zirconia Capped Piston Predicted Temperatures (Comparison of Full and Partial Insulation)

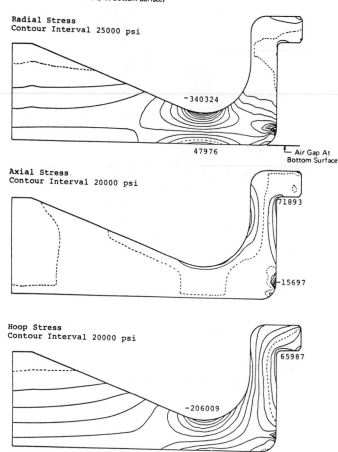

Figure 15 - NH Cap Configuration One - Thermal Case 1 Plus Assembly Interference (6-6-6) Stress Condition (Bottom Surface Free, Air Gap at Bottom Surface)

Radial Stress
Contour Interval 25000 psi

-340324

47976

Air Gap At Bottom Surface

Axial Stress
Contour Interval 20000 psi

71893

-15697

Hoop Stress
Contour Interval 20000 psi

65987

-206009

Table 5 – Comparison of Mechanical Losses Which are Reduced in MFE (in psi at 1900 rpm)

	Adiabatic Turbocompound Firing	MFE Firing	% BSFC Improvement
Piston/Rings	7.29	2.03	2.1
Con Rod and Crankshaft Bearings	4.34	.83	1.4
Oil Pump	1.50	0	.6
Total			4.1

Stress analysis using finite element modelling indicates that the low thermal conductivity, high thermal expansion material properties result in high thermal stresses, as shown in Figure 15. The critical stresses are also dependent on the thermal boundary conditions and the geometric configuration of the assembly. Currently, development work is being conducted on this type of approach.

There appears to be much room remaining in material development and design optimization in order to reduce stresses and further improve component reliability.

ADVANCED MINIMUM FRICTION ENGINE

The next step after the adiabatic engine concept is the minimum friction engine (MFE) concept. In this engine concept, it is hoped to reduce the mechanical friction of the engine by 50%. The engine friction contributed by the various engine components is shown in Table 5.

By using the following engine friction reduction technique, it is hoped that our target of 50% reduction can be achieved:
- oilless engine operation
- gas lubricated piston and cylinder liner
- ceramic main, crank, and wrist pin bearings
- solid lubricant gears, rocker arm bearings, etc.

A cross-sectional diagram of the adiabatic engine with the minimum friction concepts is shown in Figure 16. Figure 17 shows the NC132 ceramic bearings for the mains. The friction of the hydrodynamically lubricated crankshaft, for example, is about 8.2 fhp at rated condition. When ceramic bearings are used, this number can be reduced to less than 2.0 fhp.

29

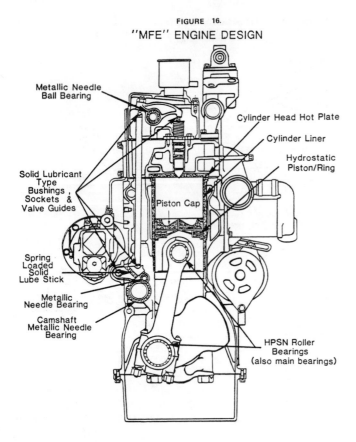

FIGURE 16.
"MFE" ENGINE DESIGN

Metallic Needle
Ball Bearing

Solid Lubricant
Type
Bushings,
Sockets &
Valve Guides

Spring
Loaded
Solid
Lube Stick

Metallic
Needle Bearing

Camshaft
Metallic Needle
Bearing

Cylinder Head Hot Plate

Cylinder Liner

Hydrostatic
Piston/Ring

Piston Cap

HPSN Roller
Bearings
(also main bearings)

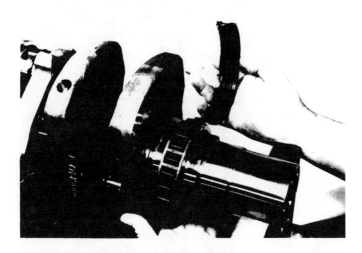

FIGURE 17.
NC-132 HPSN MAIN BEARINGS

Again, the need for high technology ceramics has been demonstrated for the advanced heat engine. The properties of ceramics which are essential for the MFE application are:
- low coefficient of expansion
- high Hertz stress capability
- high temperature and strength
- low friction property
- low wear rate

The solid lubricant plays an essential role in an oilless engine. Applications would be where relative motion between two mating parts occur with:
- marginal lubrication
- lack of lubrication
- very high temperature environment
- rocking motion
- gears

Some of the common solid lubricants investigated in the oilless adiabatic engine are shown in Table 6. Compatible low wear rates as well as low friction coefficients are desired for an acceptable solid lubricant.

Table 6 – Friction and Wear Data of Some Candidate Materials Against Cr_2O_3 Coated Rollers at 380°C

Material	Wear Rate μ/h	Friction Coefficient
Hard Chrome Plate	225	0.54
Iron/Graphite Powder Compact	134	0.43
Martensitic SG Iron (piston ring)	69	0.45
Steel Bonded TiC	9.6	0.52
M2 with LiF+Cu in Pockets	6.4	0.10-0.31
M2 with LiF+Cu in Pockets	5.7	0.12-0.22 (540°C)
Stainless Steel/ Tribaloy Powder Compact	4.6	0.51
Borided M2	1.5	0.59
Tribaloy T100 (Plasma Sprayed)	1.0	0.53
M2 with CaF_2 in Pockets	0.8	0.31
Nitrided M2	0.4	0.45
M2 with MoS_2 in Pockets	0.4	0.25
M2	0.3	0.45
Metco 505 Mo Alloy (Plasma Sprayed)	< 0.14	0.78
Chrome Oxide (Plasma Sprayed)	< 0.14	0.26
Chrome Oxide with LiF Coating (Slurry)	< 0.14	0.26

CERAMICS IN TURBOCHARGERS

In any heat engine destined for the future, turbomachines will become an essential component. High performance ceramics could play an essential role in this important engine component. Not only the aerodynamic design can be improved because of its low coefficient of expansion, but also the response characteristics because of its light weight. Variable geometry devices made from ceramic can further improve the response characteristics. The turbocharger is usually lubricated from the engine crankcase oil. For cold weather starts, maintenance free seal leakage, close clearance operation, and elimination of lubricating oil with the hydrodynamic lubrication is highly desirable. The Cummins/TACOM advanced turbocharger with Si_3N_4 turbine rotor and ceramic antifirction bearing to provide these

attributes is shown in Figure 18 and 19, respectively.
Figure 19A schematically shows a Cummins T-46
turbo fitted with a ceramic rotor and ceramic bearings.

FIGURE 18 - PRESSURELESS SINTERED SILICON NITRIDE
TURBINE ROTOR FOR TURBOCHARGER

FIGURE 19 - NC 132 HPSN SILICON NITRIDE BALL BEARINGS
FOR TURBOCHARGER

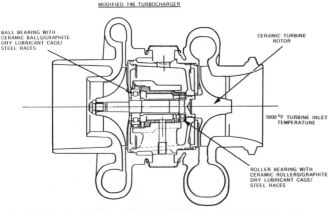

FIGURE 19A - SCHEMATIC OF MODIFIED T46 TURBOCHARGER
WITH SILICON NITRIDE TURBINE ROTOR AND
BEARINGS

Ceramics offer significant advantages over the
metal counterpart if the ceramic properties are used
properly. Thus far, the turbine rotors appear to be a
metal copy. Again, the very properties that make
ceramics valuable in turbocharger applications are:

. low coefficient of expansion
. high strength and temperature capability
. light weight for response characteristics
. lack of strategic materials

UNCOOLED FIRST GENERATION ADIABATIC ENGINE

The duplex coating developed at Cummins for
engine application was quite successful. It was
decided to fabricate a first generation adiabatic
engine and install it on a U.S. Army five-ton truck to
assess its overall performance on the road. The front
end of the five-ton truck was shortened one foot to
demonstrate its compactness when installed without a
radiator, fan, water pump, and other ancilliary
equipment. Figure 20 shows the front end of the truck
with the hood open to show its compact installation.
The now superfluous accessories and equipment not
needed on the truck are shown in Figure 21. The
passenger compartment is heated by the 265°F
lubricating oil. The main advantages of the truck are
illustrated by the following:

Size Reduction	: 20 ft^3
Weight Reduction	: 338 pounds
Number of Parts	
Eliminated	: 361 Parts
Coolant	: 42 quarts

The engine that was installed in the truck is
shown in Figure 22. The truck performed quite well
over the road from Detroit to Washington, D.C. The
performance of the engine is shown in Figure 23.
Without any cooling system parasitics in an installed
vehicle, the truck delivered a respectable 9.2 miles
per gallon.

AA750

The first generation adiabatic engine was
fabricated, assembled, engine tested, and also road
tested as described above. The ceramic components
for the prototype 5-1/2 x 6 single engine was also
fabricated and tested in a single cylinder engine with
specifications described in "Operating Environment".
Upon successful demonstration of the prototype
adiabatic engine, an advanced adiabatic AA750 engine
was started. A prototype demonstration is schedule
for mid-1984.
The AA750 will come in two builds. A
turbocharged 600 HP and a turbocompound 700 HP are
scheduled on a Cummins V903 engine block. The
specifications of the AA750 are as follows:

Bore x Storke, inches	5-1/2 x 4-3/4
Engine Speed, rpm	3200
Displacement, in^3	903
BMEP	
Minimum BSFC, lb/bhp-hr	0.28
Boost Pressure Ratio	2.87
Horsepower	600 (turbocharged)
	700 (turbocompounded)
Engine Weight, lb	2590

A cross section of the AA750 engine is shown in
Figure 24. The exhaust and inlet port insulation is
cast en bloc aluminum titanate (TIALIT). Cylinder
head and piston insulation is achieved by partially
stabilized zirconia. The cylinder liner is chrome oxide

FIGURE 20 - UNCOOLED ADIABATIC ENGINE

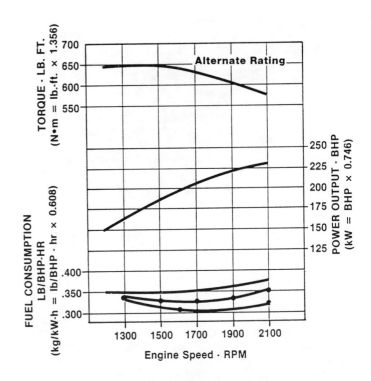

FIGURE 23 - PERFORMANCE CURVE OF THE UNCOOLED
250 HP 5 - TON TRUCK ENGINE

FIGURE 21 - ELIMINATED PRODUCTION HARDWARE
- UNCOOLED ADIABATIC ENGINE

FIGURE 22 - UNCOOLED 250 HP FIRST GENERATION
ADIABATIC DIESEL ENGINE FOR THE
5 - TON TRUCK

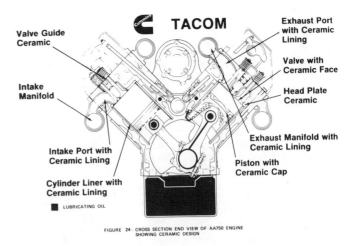

FIGURE 24 - CROSS SECTION END VIEW OF AA750 ENGINE
SHOWING CERAMIC DESIGN

coating over plasma sprayed zirconia. The turbomachine for the engine is expected to have an overall efficiency of 0.64. Overall efficiency is the product of compressor efficiency, turbine efficiency, and mechanical efficiency.

CONCLUSIONS

The feasibility of an adiabatic turbocompound engine has been demonstrated. Much work remains to be done before the adiabatic engine can be considered a commercial reality. The success of the adiabatic concept and other future advanced heat engine concepts will depend on the ability of the ceramic industry meeting the requirements of these power plants. Indeed the future appears bright and rewarding to the engine manufacturers and the ceramic industry. It could be costly to the laggards. In conclusion:

. The feasibility of a ceramic adiabatic engine has been demonstrated.
. A waste energy conversion device is necessary to take full advantage of the adiabatic diesel engine.
. The advanced yttria stabilized partially stabilized zirconia has demonstrated its ability to meet most of the adiabatic engine design criteria.
. The ceramic coatings offer an immediate alternative solution to the monolithic design approach and could possibly be the first commercial engine on the market.
. High technology ceramics offer an opportunity to the engine manufacturer to consider a waterless oilless, and high efficiency power plant for the future. Ceramics appear applicable in many engine components and accessories as well.

The use of ceramics in engines with all its associated benefits will not come easy. The problems that need to be resolved are:
A. High temperature lubrication
B. High temperature, high strength, and insulative materials
C. Ceramic bearings
D. Gas lubricated bearings
E. Low cost fabrication
F. Low cost finishing and machining
G. Solid lubricants
H. Better ceramic coatings

Of the above problem areas, A, E, and F will probably present the greatest challenge.

Higher temperature, higher pressure thermodynamic cycles will be the way for heat engines of the future. Since these temperatures and pressures are beyond the current limit of metal, high strength, high temperature ceramics will be the material to reckon with in the future. The manufacturing technology, evaluation technology, and the application technology of ceramic materials are many; but they appear to be tenable.

The ceramic technology has been moving at a rapid pace in recent years. In many ceramic materials, expansion coefficients, conductivity, etc. can be varied to meet the needs of the applications. Some of the current favorite monolithic ceramics and coatings which have satisfactorily met the needs of the adiabatic diesels were presented. There is bound to be more new materials in the future. However, for the present, the materials which have met the

requirements of an adiabatic engine for various adiabatic diesel components are summarized in Table 7.

TABLE 7 - SUMMARY OF HIGH TECHNOLOGY CERAMICS FOR ADIABATIC ENGINES IN JAPAN & U.S.A.

ADIABATIC COMPONENTS	LOW FRICTION	LIGHT WEIGHT	INSULATION	WEAR RESISTANCE	HEAT RESISTANCE	CORROSION RESISTANCE	EXPANSION COEFFICIENT	HIGH TECHNOLOGY CERAMICS
PISTON		•	•		•	•	•	Si_3N_4, PSZ, TTA
PISTON RING				•	•	•		SSN, PSZ, COATING
CYLINDER LINER	•		•		•	•		Si_3N_4, PSZ, COATING
PRECHAMBER			•		•	•		PSZ, Si_3N4
VALVE		•		•	•	•		SSN, PSZ, COMPOSITE
VALVE SEAT INSERT				•	•	•		PSZ, SSN
VALVE GUIDES	•			•	•	•		PSZ, SSN, SiC
EXHAUST / INTAKE PORTS			•		•	•		ZrO_2, Si_3N_4, $TiO_2Al_2O_3$
MANIFOLDS			•		•	•		ZrO_2, Si_3N_4, $TiO_2Al_2O_3$
TAPPETS		•		•	•			PSZ, SiC, Si_3N_4
MECHANICAL SEALS	•			•	•			SiC, Si_3N_4, PSZ
TURBOCHARGER								
TURBINE ROTOR		•	•		•	•	•	Si_3N_4, SiC
TURBINE HOUSING			•		•	•	•	LAS
HEAT SHIELD			•		•	•	•	Z_2O_2, LAS
CERAMIC BEARINGS	•	•		•	•	•	•	SSN

NOMENCLATURE :	SSN - SINTERED SILICON NITRIDE PSZ - PARTIALLY STABILIZED ZIRCONIA LAS - LITHIUM ALUMINA SILICATE TTA - TRANSFORMATION TOUGHENED ALUMINA

ACKNOWLEDGEMENT

The authors wish to acknowledge the following for making this paper possible: Cummins Engine Company and the Advanced Engines and Systems Team, the U.S. Army Tank-Automotive Command, and Kaman Sciences Corporation.

REFERENCES

1. R. Kamo and W. Bryzik, "Adiabatic Turbocompound Diesel", 15th International Congress on Combustion Engines, June, 1983, Paris, France, pp. 417-456.

2. R. Kamo and W. Bryzik, "Ceramics for Adiabatic Turbocompound Engine", Proc. of Sixth Army Material Technology Conference on Ceramics for High Performance Applications, III, 1979, Orcas Island, Washington, pp. 187-216.

3. W. Bryzik and R. Kamo, "TACOM/Cummins Adiabatic Engine Program", SAE Paper 830314, The Adiabatic Diesel Engine SP-543, pp. 21-45, SAE International Congress & Exposition, Detroit, MI, 1983.

4. R. Kamo, M. E. Woods, and W. C. Geary, "Ceramics for Adiabatic Diesel Engines", CIMTECH 4th, St. Vincent, Italy, June, 1979.

5. M. E. Woods and I. Oda, "PSZ Ceramics for Adiabatic Engine Components:, SAE Paper 820429, February, 1982, Detroit, MI.

6. J. I. Mueller, A. S. Kobayashi, and W. D. Scott, "Designs with Brittle Materials", University of Washington, Seattle, 1979.

7. S. Yamamoto and I. Oda, ZIRCOA 83, Stuttgart, Germany, June, 1983.

8. S. Timoney and G. Flynn, "A Low Friction Unlubricated SiC Diesel Engine", SAE Paper 830313, The Adiabatic Engine SP-543, pp. 11-19, Detroit, MI, 1983.

9. R. A. Miller and C. A. Lowell, "Failure Mechanisms of Thermal Barrier Coatings Exposed to Elevated Temperatures" International Conference on Metallurgical Coatings and Process Technology, San Diego, CA, April, 1982.

10. S. Stecura, "Effect of Plasma Spray Parameters on Two Layer Thermal Barrier Coating System Life", NASA Technical Memorandum 81724, March, 1981.

11. R. Kamo and W. Bryzik, "Ceramics for Adiabatic Engine", La Societe Des Ingenieurs De L'Automobile, "Les Materiaux Dans L'Evolution Des Moteurs", November-December, 1982, Paris, France.

12. J. H. Stang, "Designing Adiabatic Engine Components", SAE Paper 780069, February, 1978, Detroit, MI.

13. R. Kamo, M. E. Woods, and J. Jones, "Use of Ceramics for Adiabatic Diesel Engine", 6th International Symposium on Ceramics, Bologna, Italy, September, 1983.

840429

Tribology at High Temperature for Uncooled Heat Insulated Engine

**Tomoki Shimauchi, Taku Murakami
and Tsutomu Nakagaki**
Komatsu Ltd. (Japan)
Yuko Tsuya and Kazunori Umeda
Mechanical Engineering Laboratory (Japan)

ABSTRACT

Preliminary study of a heat insulated diesel engine demanded more than 350°C operating temperature of the cylinder liner temperature at top ring reversal. In order to define the tribological problems of the engine, a single cylinder endurance run was conducted and friction loss of the six cylinder heat insulated engine under operating condition was measured.

A major tribological problem of piston ring and liner interface was found to be choosing proper wear resisting coatings for their surfaces as the hydrodynamic film protection at top ring reversal reduced significantly.

Considering possible applications to the piston ring and cylinder liner sliding faces, several ceramic coating materials were evaluated for their load carrying capacities and wear resisting properties by laboratory wear test. The results showed that the sliding performances of the coating materials changed considerablly depending on the combination of materials and lubricants.

EVER SINCE the beginning of the price rise of petrolium fuel, development of high thermal efficiency diesel engines was a major concern of engine manufacturers of the world. A new engine concept, an adiabatic turbocompound engine was proposed by Kamo and attracted attention for its potential high thermal efficiency and other advantages, such as low maintenance cost and higher reliability as a result of elimination of radiater cooling system (1)(2)*. With the use of high temperature heat resisting ceramics, the engine can be operated without coolant and the heat normally dissipated to the coolant goes to exhaust energy, which can be recovered as shaft output through power recovery system.

The authors organization has conducted preliminary studies on the heat insulated turbocompound engine and showed its excellent advantages with respect to fuel consumption and applicability for low quality fuel (3)(4). This paper describes some of the tribological problems of the engine and results of laboratory wear test of several ceramic coating materials for the ring and liner surfaces.

ENGINE TESTS

Both single cylinder and six cylinder engine tests were conducted, the former mainly for preliminary reliability evaluation of the heat insulating components and the later mainly for performance evaluation of the engine. The specifications of the engine used as a base for heat insulation are shown in Table 1. Figure 1 is a sketch of the cross section of the modified engine for heat insulation which is described in more detail in reference (3). Single cylinder endurance test reported here was conducted using components of a larger bore version of the base engine (110 mm bore, 125 mm stroke).

Single cylinder endurance test run.
- After several trial test of the heat insulating components, 250 hours endur-

*Numbers in parentheses designate References at end of paper

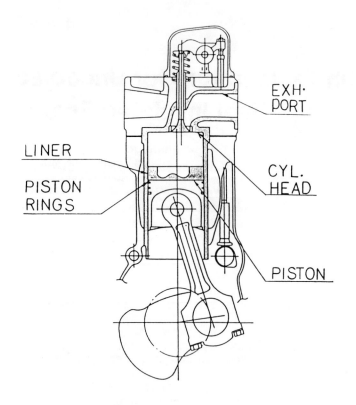

FIG 1. SKETCH OF THE UNCOOLED HEAT INSULATED ENGINE

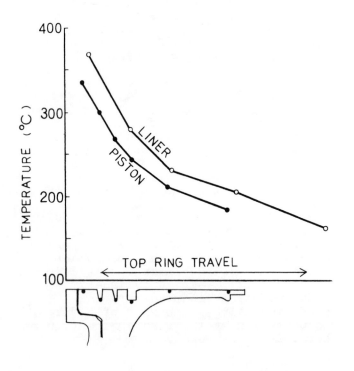

FIG 2. PISTON AND LINER TEMPERATURES UNDER SINGLE CYLINDER
ENDURANCE TEST CONDITION.

TABLE 1. - BASE ENGINE SPECIFICATIONS

ENGINE TYPE	
COMBUSTION	DIRECT INJECTION
ASPIRATION	TURBOCHARGED
COOLING	WATER COOLED
BORE	105 mm
STROKE	125 mm
DISPLACEMENT	6.5 liter
RATED POWER	120 kw
RATED SPEED	2500 rpm

ance test run was performed by the single
cylinder engine. The test conditions
were chosen as given in Table 2 which
assumes heavy load condition of a turbocompounded engine. Temperatures of the
piston and the cylinder liner were
measured under this condition and shown
in Figure 2. The piston was a ceramic
composite type comprising a ceramic
crown and an aluminum skirt bolted
together. Table 3 shows the ring
package and the cylinder liner facing
used for the endurance run. The
lubricant was mineral oil of the
viscosity number SAE40 which was
formulated for high temperature
stability (D in table 5)

TABLE 2. - TEST CONDITION OF THE SINGLE
CYLINDER ENDURANCE RUN

REVOLUTION SPEED	2500 rpm
FUELLING RATE	7.8 bar
INTAKE PRESSURE	3.3 bar
EXHAUST PRESSURE	4.0 bar
INTAKE TEMPERATURE	100 °C
OIL PAN TEMPERATURE	120 °C
OIL PAN CAPACITY	8 liter

TABLE 3.   RINGS AND LINER FOR ENDURANCE RUN

TOP RING	TITANIUM CARBIDE PLASMA SPRAYED
	BARREL FACE
	KEYSTONE TYPE
2ND RING	TITANIUM CARBIDE PLASMA SPRAYED
	TAPER FACE
	HALF KEYSTONE TYPE
OIL RING	CHROME PLATED
	COIL EXPANDER TYPE
CYL. LINER	CHROMIUM CARBIDE PLASMA SPRAYED

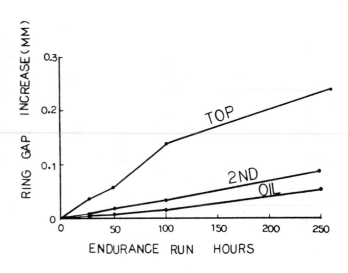

FIG 3.   PISTON RING WEAR

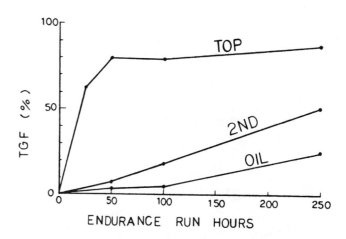

FIG 4.   RING GROOVE CARBON FILLING (TGF)

Results of the endurance test run.
- During the test, the engine was dis-
assembled several times for the
inspection of piston ring wear and ring
groove carbon filling which are shown in
Figure 3 and Figure 4. Ring wear was
found to increase almost in proportion
to the endurance time. Top ring groove
carbon filling increased rapidly at the
beginning stage and stabilized at about
80 % for the rest of the endurance run.
This may be considered as a self
cleaning effect of a key stone type
piston ring. Oil consumption and blow-
by were continuously monitored which are
shown in Figure 5. Both oil consumption
and blow-by were high at the beginning
but stabilized after about 70 hours.
Oil samples were taken on 25 hours
interval for analysis and the results
are shown in Figure 6. Noticeable
increase of viscosity indicated
evaporation loss of the lighter fraction
of the oil. After the test, cylinder
liner side wear at top ring reversal
position was measured by profile chart
as shown in Figure 7 and the average
wear was 0.018 mm. The piston after
the test is shown in Figure 8. Although
carbon and lacquer deposits were heavy
compared with the water cooled engine,
carbon on the top land and in the top
ring groove was found to be soft which
easily came off.

Six cylinder engine test - Friction
loss of both the heat insulated and the
water cooled base engine were measured
by DC dynamometer. A conventional
method which measures the steady state
motoring torque under controlled coolant
temperature could not be used for the
uncooled heat insulated engine.

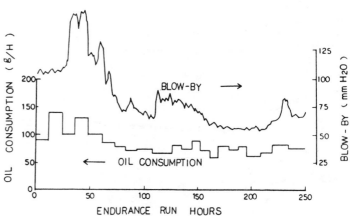

FIG 5.   OIL CONSUMPTION AND BLOW-BY

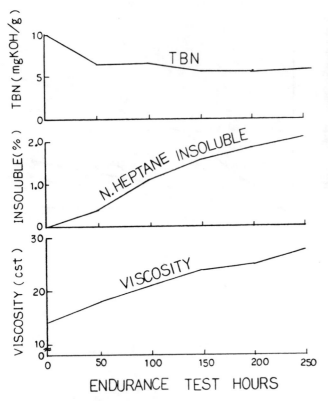

FIG 6.   LUBRICANT PROPERTIES CHANGE

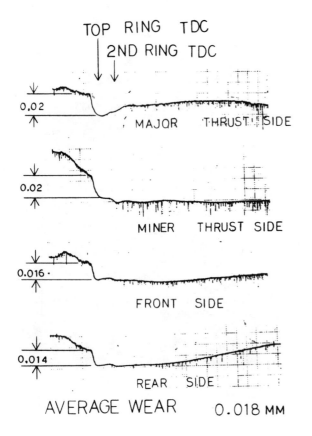

FIG 7.   CYLINDER LINER SIDE WEAR

FIG 8.   PISTON AFTER THE ENDURANCE TEST

Estimation of the friction loss of
the engine was made as follows. The
engine was operated under certain load
while the dynamometer was under constant
speed control. When the condition
became steady the fuel supply to the
engine was instantaneously cut off and
the dynamometer automatically switched
from power absorption to motoring. This
motoring torque was recorded on a pen
chart recorder until it stabilized
to steady state as shown in Figure 9.
The friction loss at the instance of the
fuel cut off was estimated by
extraporation as a dot shown in the
figure. The pressure difference
between intake and exhaust manifold is
also shown in the figure. This was due
to thermal and mechanical inertia of the
turbo-charging system. The extrapolated
instantaneous friction loss contained
pumping work by this pressure difference
which varied with the engine load at the
furl cut off. In order to correct this
effect, $\triangle p$ in the figure was subtracted
from the instantaneous friction mean
effective pressure (FMEP). By this
method the corrected FMEP was considered
to represent operating condition except
that the effect of the firing pressure
of the expansion stroke was not included.
The same procedure was applied on the
water cooled base engine and the results
are shown in Figure 10. In the figure,
measured temperatures of the cylinder
liners at top ring reversal position are
included. Corrected FMEP of the water
cooled engine was almost constant with
engine load, which justified the
conventional steady state motoring
method. But corrected FMEP of the heat
insulated engine decreased as the engine
load increased. This suggested that
even under this high temperature
operation, the tribological condition of
the piston ring and liner interface was

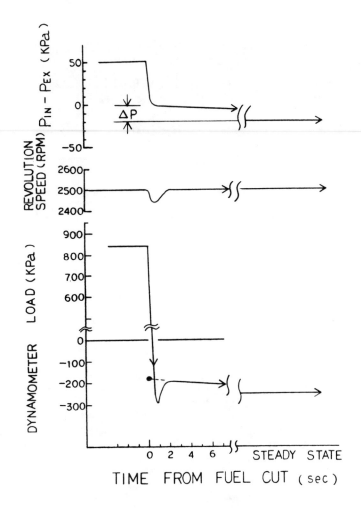

FIG 9.   MEASUREMENT OF THE INSTANTANEOUS FRICTION LOSS
OF THE ENGINE (LOAD IS EXPRESSED IN ENGINE MEAN
EFFECTIVE PRESSURE).

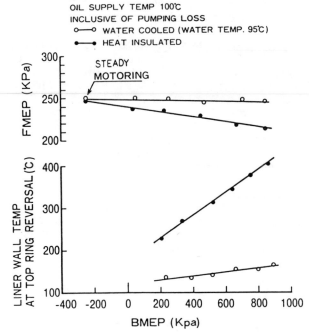

FIG 10.   COMPARISON OF FRICTION MEAN EFFECTIVE PRESSURE

mainly hydrodynamic, though high side
wear rate of the cylinder liner shown
in Figure 8 indicated severe boundary
lubrication condition near the top ring
reversal position of the firing stroke.
    Thus the engine test results
indicated that one of the problems to
be solved was to reduce the wear of the
top ring reversal contact under high
temperature condition.

LABORATORY FRICTION TEST

    Heat insulated engine required high
temperature operation of the cylinder
liner interface. Ceramic coatings were
considered to be possible sliding
materials for the engine. In order to
understand their tribological
characteristics, evaluation test of the
several candidate materials were
conducted by a Falex tester of Figure 11.
The maximum temperature was limited to

FIG 11.   FALEX TESTER

300°C because of caution against fire. This is not as high as the operating temperature of the engine, but considered adequate for the present study.

Materials tested - Coating materials listed in Table 4 were applied mainly by plasma spray method on the sliding surfaces of the Falex test pieces which are shown in Figure 12. Thickness of the coatings was 0.2mm. Lubricants used for the test are listed in Table 5. Lubricant A is a low viscosity oil which contains sulphur and phosphorus additives. Lubricant B is ester base oil which contains no additives and C is formulated ester base synthetic engine oil which contains ZnDTP and calcium detergent dispersant. Viscosity at 300°C in the table were extrapolated from the data at 40°C and 100°C.

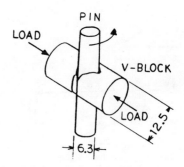

REVOLUTION SPEED   290 ±10 rpm
SLIDING VELOCITY   0.096 m/s

FIG 12.   FALEX TEST PIECES

TABLE 4.   MATERIAL TABLE

No	Base	Main Composition
1	Cr_2O_3	Cr_2O_3 (90%)
2	Cr_2O_3	Cr_2O_3, SiO_2
3	Cr_2O_3	Cr_2O_3, Al_2O_3, SiO_2
4	CrC	CrC, Ni, Cr
5	TiO_2	TiO_2 (95%)
6	TiO_2	TiO_2, Al_2O_3
7	WC	WC, CO
8	Al_2O_3	Al_2O_3, TiO_3
9	Al_2O_3	Al_2O_3
10	ZrO_2	ZrO_2, Y_2O_3
11	-	VC (TD-Process)
12	Cr_2O_3	Cr_2O_3, NbO_2 (10%)
13	Cr_2O_3	Cr_2O_3, NbO_2 (50%)
14	Cr_2O_3	Cr_2O_3, VC (10%)
15	Cr_2O_3	Cr_2O_3, VC (20%)
16	Cr_2O_3	Cr_2O_3, VC (40%)

Test procedure - In order to evaluate load carrying capacity of the coatings, load was applied stepwise by 225N untill seizure occured or up to 3600N. Each load step was kept constant for one minute and friction coefficient was calculated based on the measured friction torque. Anti-wear performance of the materials were evaluated under constant load test. The load was kept 675N for 60 minutes at 300°C. Comparison was made by wear volume calculated based on the surface profile chart of the worn track of the test pieces. Drops of lubricant was fed by a syringe to the sliding part of a test piece at about 0.01 cm³ per minute. The test pieces were heated by blowing hot nitrogen gas through a heater tube. The temperature of the test pieces was measured by a infrared thermometer and the heater power was controlled so that the measured temperature was at 300°C.

Test results of like material pairs - Figure 13 and Figure 14 are summeries of test results of each coating materials. Lubricants used and test temperatures are shown in each figures. Steel, when lubricated by low viscosity oil A at room temperature, did not seize up to 3600N, but seized easily at low load when the temperature was raised to 300°C. When lubricated by oil B or C steel did not seize even when the temperature was 300°C. In comparison with this, Cr_2O_3 did not seize at 300°C when lubricated by any of the three oils. Other materials showed different characteristics depending on the combination with the lubricants such that TiO_2 had good load carrying

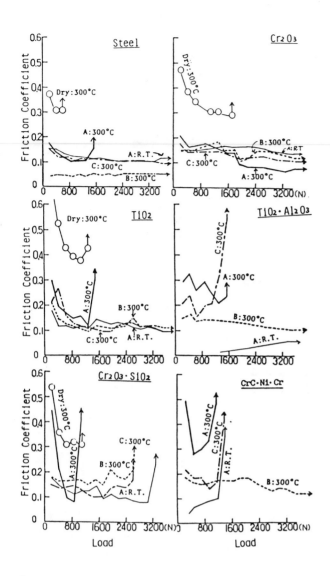

FIG 13. FRICTION CHARACTERISTICS OF COATING MATERIALS

AS LIKE PAIRS.

capacity with oil B and C but $TiO_2 \cdot Al_2O_3$ showed poor load carrying capacity with oil C. In order to show the effects of lubricant more clearly the data at 300°C were rearranged in accordance with the lubricants and shown in Figure 14. When the lubricant was oil A or B, only VC, Cr_2O_3 and $Cr_2O_3 \cdot Al_2O_3 \cdot SiO_3$ survived, but most materials performed well when the oil was B.

Under dry condition without oil, all the materials tested showed high friction coefficient at start and seized at low load after showing the tendency of a moderate decrease of friction coefficient. Among them Cr_2O_3 performed best for its dry load carrying capacity.

In the figures shown above Al_2O_3 and ZrO_2 are not included. This is because even when they were lubricated by oil A they seized at less than 675N, seizure load of dry steel, and further tests on these materials were not tried.

Figure 15 shows the effects of addition of NbO_2 and VC to Cr_2O_3. Under dry condition all the materials tested showed similar results, but when lubricated by oil A all the materials performed poorer than the base material Cr_2O_3.

In order to see the effect of load increasing procedure, load step was increased from 225N to 1125N per minute and the results are shown in Figure 16. The test material was Cr_2O_3 and the test temperature was 300°C. When the oil was A or C seizure occured at the first load step while the seizure load of oil B was

TABLE 5.　　　　OIL TABLE

		Viscosity (cst)			Flash Point	Chemical Analysis (%)		
		40°	100°	300°	(°C)	S	P	Zn
A	Highly Refined Synthetic Oil EP-aditive (SP type)	7.99	2.26	0.41	162	1.35	0.07	-
B	Synthetic Oil (ester base)	48.91	9.57	1.21	294	-	-	-
C	Synthetic Engine Oil (SD grade, ester base)	120.2	15.14	1.32	242	0.44	0.06	0.06
D	Formulated Engine Oil (Mineral base)	147.0	14.3	1.1	252	-	-	-

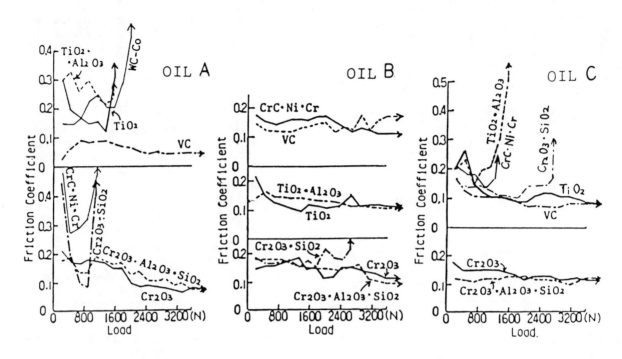

FIG 14.   FRICTION CHARACTERISTICS OF COATING MATERIALS AS LIKE PAIRS.
                        (EFFECTS OF DIFFERENT OILS)

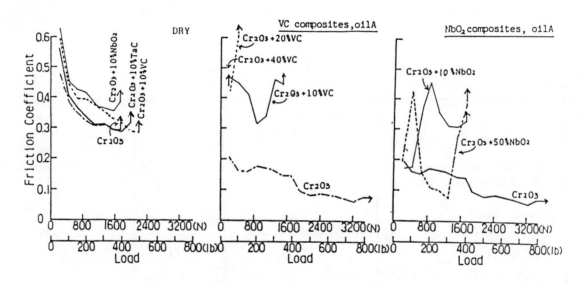

FIG 15.   FRICTION CHARACTERISTICS OF Cr_2O_3 BASE MATERIALS.
                        (EFFECTS OF VC OR NbO_2 ADDITION)

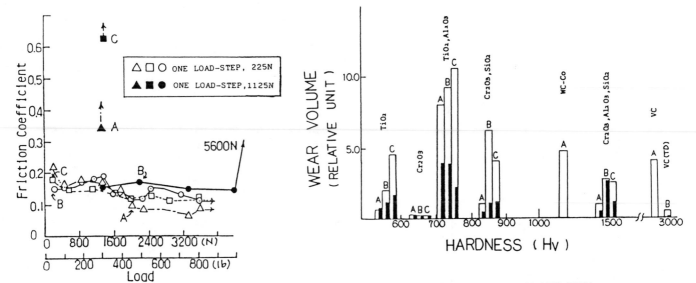

FIG 16. FRICTION CHARACTERISTICS OF Cr_2O_3
AS LIKE PAIRS
(EFFECTS OF DIFFERENT LOAD STEP)

FIG 17. COMPARISON OF WEAR AS LIKE PAIRS

as high as 5600N. This suggested that different lubrication mechanism was working between oil B and oil A or C.

Wear test results.- Wear was evaluated by volume wear measured by the surface profile chart of the worn track of the test pieces and results are shown in Figure 17. The abscissa is the vickers hardness of the coatings and the ordinate is the wear volume in relative unit. Wears of block pieces are shown by white bars and wears of pins are shown by solid bars. Noticeably small wear was found in Cr_2O_3 while the wear of $TiO_2 Al_2O_3$ was the

biggest. The worn surfaces of the pin sliding track were observed by a microscope and shown in Figure 18 with surface roughness charts. Cr_2O_3 showed the smoothest worn surface while $TiO_2 \cdot Al_2O_3$ showed the roughest surface.

The worn surfaces of these pins were also analyzed by electron probe micro analyzer (EPMA) to compare the surface atomic concentration change by friction. The results were summarized in Figure 19 in which ratios of EPMA peak heights of sulphur, phosphate, oxygen and carbon atoms of worn track surfaces to those of unworn surfaces were shown. Cr_2O_3 showed greater increase of S and P than TiO_2.

Test results of unlike material pairs.- In order to see the effects of material combinations on sliding

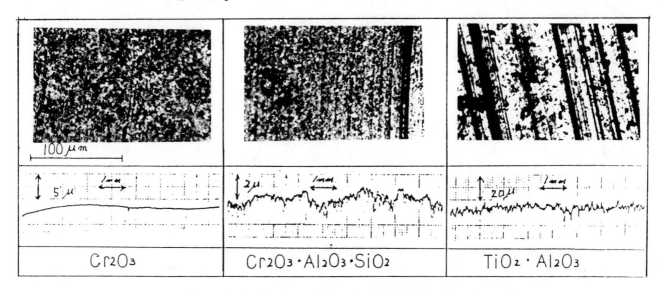

FIG 18. OBSERVATION OF WORN SURFACES

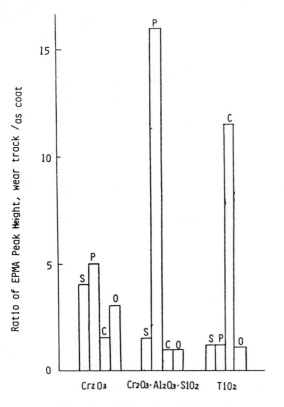

FIG 19. SURFACE ATOMIC CONCENTRATION CHANGE
BY WEAR

characteristics, test of unlike material pairs were conducted. Cr_2O_3 Al_2O_3 SiO_2, WC Co and metal Mo were chosen as key materials. The oil used was B and the temperature was 300°C. Figure 20(a1) (a2) (a3) shows the test results in which one sliding surfaces were Cr_2O_3 containing materials. When they were tested as like material pairs they did not seize up to 3600N, but when they were tested against unlike materials the sliding characteristics changed considerablly. When the counterparts were metal molybdenum or cermet (WC-Co) noticeable deterioration of load carrying capacity was found. In contrast to this, when they were tested against Cr_2O_3 SiO_2 or TiO_2 they did not seize similar to the results of like material pairs.

When one sliding part was TiO_2, combination with Mo gave poorer results than as like material pairs while no deteriorating effects were found with WC-Co. Combination of Mo and WC-Co did not seize up to 3600N. These are shown in Figure 20 (c),(d).

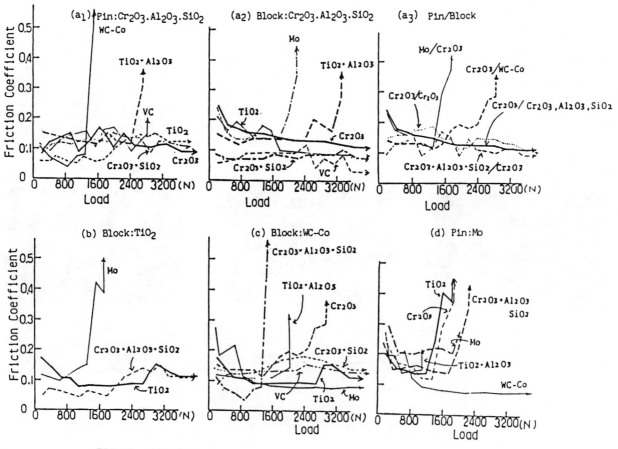

FIG 20. FRICTION CHARACTERISTICS AS UNLIKE MATERIAL PAIRS

CONCLUSIONS

1.   Tribological condition of the piston ring and liner interface of the uncooled heat insulated diesel engine prooved to be mainly hydrodynamic except at the position of top ring reversal.

2.   High top ring reversal temperature of the heat insulated engine caused the severe side wear of the liner at top reversal position.

3.   Laboratory sliding test showed that the sliding characteristics of the ceramic coating materials changed considerablly, depending on the combination of materials and lubricant. Careful attention for operating environment is important before actual application of ceramic coatings to engine components.

4.   Under dry condition all the ceramic materials tested seized at very low load. Direct application of ceramic materials into engine components under dry condition appears to be difficult.

REFERENCES

1.   R. Kamo and W. Bryzik, "Adiabatic Turbocompound Engine Performance Prediction", SAE Paper 780068, 1978.

2.   R. Kamo and W. Bryzik, "Cummins-TARADCOM Adiabatic Turbo-compound Engine Program". SAE Paper 810070, 1981.

3.   T. Yoshimitsu et.al., "Capabilities of Heat Insulated Diesel Engine "SAE Paper 820431, Detroit Mich., Feb. 1982.

4.   K. Toyama et.al., "Heat Insulated Turbocompound Engine" SAE Paper 831345, 1983.

Assessment of Positive Displacement Supercharging and Compounding of Adiabatic Diesel

Jaroslav Wurm, John A. Kinast
and Tytus Bulicz
Space Conditioning Research
Institute of Gas Technology

ABSTRACT

A new alternative to turbocompounding for converting exhaust heat from adiabatic diesels to mechanical energy is evaluated. The concept is based on a positive displacement screw compressor and expander pair applied to an adiabatic diesel for supercharging and mechanical power compounding to form a compact and efficient power system. Predictions of theoretical performance for the system and its components are given, and the technical feasibility and potential advantages of this system are discussed.

THE TECHNICAL DEVELOPMENTS THAT HAVE OCCURRED RECENTLY have consistently proven the advantages of a diesel engine as both an automotive and as a stationary prime mover. In addition, the diesel engine offers the potential for significant improvements in performance characteristics such as fuel consumption, specific power output, and emission control. Over the last several years the promise of these potentials has led to advanced diesel R&D programs on such topics as variable area turbocharging, variable compression ratio, high pressure fuel injection systems, turbocompounding, and the advanced concepts of adiabatic engines and Rankine bottoming cycles. The combination of several of these concepts resulted in the development of an advanced compression ignition internal combustion engine. The new concept consists of an adiabatic diesel with high supercharging and some method of converting after-cylinder heat energy to mechanical power. Thermodynamically, such a system operates according to a theoretical cycle known as the Atkinson Cycle, shown in Figure 1. Until now, such a cycle was practically realized by a turbocharged/turbocompounded system, consisting of a diesel engine with two gas turbines connected in series. The first turbine drives a turbocompressor; the second one

acts as a turboexpander, converting surplus exhaust gas energy into mechanical work. Another possible approach to exhaust energy utilization for diesel engines is through the use of a positive displacement expander in place of the turboexpander. The feasibility of this approach, its theoretical and practical advantages and disadvantages over turbocompounding, and a review of the preliminary design analysis for positive displacement power compound systems are the topics of this paper.

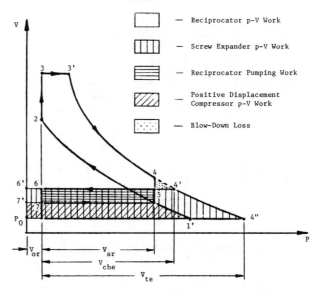

Figure 1. ATKINSON CYCLE GENERATED BY DIESEL ENGINE AND SCREW EXPANDER AND COMPRESSOR

GENERAL DESCRIPTION OF THE SYSTEM AND OPERATION

The evaluated power system consists of a mechanically supercharged adiabatic diesel engine, the exhaust of which is connected to a positive displacement expander. Both supercharger (compressor) and expander are positive displacement, double-screw machines. The use of rotary screw compressors as diesel superchargers has met with limited success in the past, due to the disadvantages of mechanical supercharging with a machine of fixed delivery characteristics. Our approach differs from that original concept, as will be shown.

The combined working cycle of the system together with the thermodynamic interactions between the components can be observed in Figure 1. The generalized layout of the system and the heat energy and mechanical energy flows are schematically shown in Figure 2. Referring to Figure 1, the surplus after-cylinder heat energy leaves the adiabatic engine and enters the screw expander at temperature T_4' and pressure p_4'. Assuming that steady-state and adiabatic conditions exist between an engine and an expander, temperature T_4' is controlled by engine operating conditions and thermodynamic cycle and back pressure p_4' is controlled both by the operating conditions of engine and volume flow characteristics of the expander. Single-stage expansion of exhaust gases from p_4' to ambient pressure p_0 by the screw expander generates the mechanical power which is coupled to the engine shaft. The positive displacement screw compressor driven by power from the engine shaft delivers the air flow required for the adiabatic diesel boost, characterized by pressure p_7.

positive displacement is provided by cyclic changes of volumes bounded by the lobes of two rotors in continuous mesh during rotation, shown schematically in Figures 3 and 4. The lines in Figure 3 represent rotor lobes while the areas between them represent the inter-lobe volumes. The charge enters the inter-lobe volume through the inlet port in the high pressure end plate of the screw expander. The size of the inter-lobe volume continuously varies with the rotation of rotors from 0 to the final value of V_{che} corresponding to the charging volume of the expander. Charging of the inter-lobe space ends when it is sealed off from the intake. As the rotors further rotate, the inter-lobe volume increases causing an expansion of the trapped gases. Expansion within the individual inter-lobe volume is completed when the inter-lobe volume opens to the exhaust port in the low pressure end plate of the expander, ideally occurring when the maximum inter-lobe volume is reached. The expanded gas is discharged as the rotors complete the cycle. Figure 4 shows the volumes and their approximate occurrence along the length of the expander. The volume characteristics for a general inter-lobe cell as a function of the male rotor rotation angle is shown in Figure 5. It can be seen that an analog exists to the piston expander with valve controlled inlet and outlet. The four-lobed male rotor expander has volume characteristics very similar to a four-cylinder two-stroke engine.

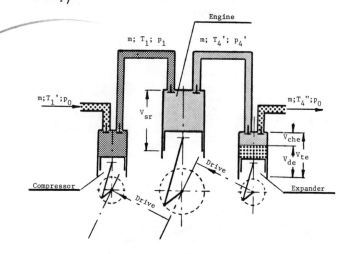

Figure 2. SCHEMATIC OF POSITIVE DISPLACEMENT COMPOUNDED DIESEL SYSTEM

Analysis of the screw expander's potential performance has shown that optimal efficiency can be obtained for operating pressure ratios from 1.5 to 4.0, approximately. This range of optimal expander pressure ratios is similar to the range of back-pressures acceptable with respect to the optimal operating conditions for a diesel engine compounded with such an expander.

PRINCIPLE OF OPERATION OF A SCREW EXPANDER

The screw expander operates similarly to a reciprocating piston-type expander. In general,

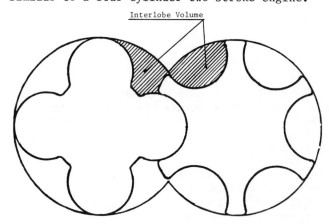

Figure 3. INTER-LOBE VOLUME OF SCREW EXPANDER

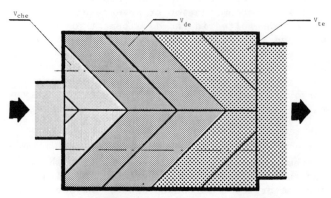

Figure 4. CHARGING, EXPANSION, AND DISCHARGE VOLUMES OF SCREW EXPANDER

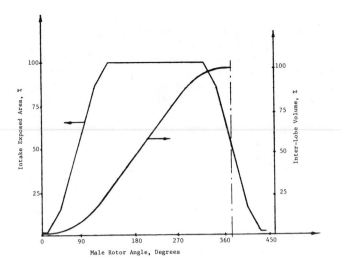

Figure 5. CHARACTERISTICS OF ONE
INTER-LOBE VOLUME

POSITIVE DISPLACEMENT SUPERCHARGING

Exhaust gas heat energy recovery by means of after-cylinder expansion inherently creates relatively high back-pressure at the exhaust of the diesel engine for the entire operating range. To decrease the potentially adverse effect that high back-pressure would have on the engine brake thermal efficiency (potential deterioration of scavenging process), air charging characteristics should ensure the existence of high pressure boost at any operating point of the system. For the high peak back-pressures potentially encountered with a compounded system, the system will require high supercharging pressure ratios between 3.0 to 4.0. Although the thermodynamics of a high-boost diesel cycle were not investigated, recent articles discussing the high pressure charging concept indicate that such a cycle is feasible and has a number of merits, especially when used with an adiabatic diesel. A review of the available types of positive displacement compressors that could be considered as compact one-stage high-pressure superchargers has revealed that, for an engine size range of about 80 — 800 hp, a screw machine could be practical. The screw compressor operates similarly to the previously discussed expander, the basic difference being in the design of inlet and outlet porting to produce the suction, compression, and discharge sequence.

SYSTEM ANALYSIS

To more thoroughly evaluate the potential performance improvement of the positive displacement supercharged compounded system over a turbocompounded system, as well as to assess the technical feasibility of such a system, following problems have been given special consideration:

- Evaluation of overall efficiency of screw expander and screw compressor using an interpolation of efficiency characteristics for existing units and a mathematical model relating this efficiency to design and operating parameters.

- Assessment of differences in the theoretical components of overall efficiencies between screw compressor and screw expander.

- Development an engineering method for selecting and sizing a screw expander for a given combustion engine.

- Evaluation of theoretically generated performance for PDSC system and comparison to measured performance of baseline turbocompound system.

The basic areas to consider when evaluating the suitability of screw machinery used as superchargers and exhaust expanders are: the levels of efficiency available in a wide range of operating conditions imposed on them by variable loads and speeds of the diesel engine; and the general trends in the efficiency for changes in operating conditions as well as in design parameters.

In general terms the overall efficiency of the screw machine can be expressed as:

$$\eta_o = \eta_{i(cor)} \times \eta_f \times \eta_m \qquad (1)$$

where:

$\eta_{i(cor)}$ = Ideal efficiency accounting for losses from the cycle related to adiabatic work, corrected for:

- Difference between actual operating external pressure ratio and internal, built-in pressure ratio of given machine.

- Pressure drops at intake and discharge of machine depending on inlet and discharge porting design and velocity of flows.

- Mass inertia due to mode of operation (compressor or expander).

η_f = Efficiency accounting for internal and external leakage losses, dependent on: the radial clearance between the rotors and the housing walls, the axial clearance between the rotors and the end plates of the housing, the clearance between the rotors, tip speed of rotors, actual pressure ratio, and dimensions of rotors.

η_m = Mechanical efficiency, which is dependent on the design of machine and, to a lesser extent, on its operating conditions.

An important difference between compressor and expander operation for a screw machine is that leakage between inter-lobe volumes during expansion can be partially recovered, as it increases the enthalpy of the gas trapped in the inter-lobe volume receiving the leakage. This, combined with the potential use of the inertia of exhaust gases, permits the expander to achieve a higher efficiency than the one theoretically attainable for a screw compressor.

Analytical models reflecting these functional relationships were developed for steady flow conditions and without accounting for the "heat-up" effect of internal leakages. Results of the computer simulation of the overall efficiency for a wide spectrum of male rotor speeds and operating (external) pressure ratios for preliminary designs of the screw compressor and screw expander are shown in Figures 6 and 7, respectively.

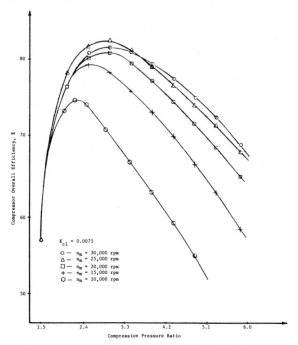

Figure 6. SCREW COMPRESSOR EFFICIENCY

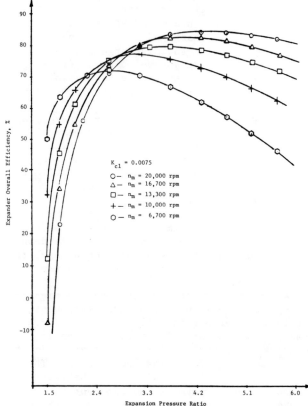

Figure 7. SCREW EXPANDER EFFICIENCY

Figure 6 shows the overall efficiency characteristics for one particular set of operating conditions of a screw compressor, with a design "built-in" pressure ratio of 3.2 and internal clearances analytically characterized by a "reduced" clearance factor, K_{cl}, of 0.0075, that was determined as a function of the external pressure ratio for several different male rotor speeds. The same type of characteristics shown in Figure 7 represents a screw expander with the design "built-in" pressure ratio of 4.3. In order to prevent the overexpansion of exhaust gases at low operating (external) pressure ratios and avoid the associated efficiency losses, the expander is assumed to be equipped with an expansion ratio modulation system. The venting losses related to operating the expansion modulation system were incorporated in the model of expander efficiency. For small differences between "built-in" and operating pressure ratios and for moderate expander speeds, this system can theoretically provide acceptable efficiency levels. However, when operating pressure ratios differ drastically from the design value at high speeds (in the model this is indicated by critical velocities of "venting" gas in the valve passages of the modulation system), efficiency would deteriorate sharply, as shown by the efficiency curve at 20,000 rpm and at low pressures interval in Figure 7. Consequently, sizing of an expander for a given engine should be based upon two essential considerations:

- The maximum back-pressure in the engine exhaust manifold developed by the expander volume flow rate capacity must be optimized with respect to the efficiency of the engine and the overall efficiency of the screw expander.

- The displacement of the expander should provide for total expansion of the working gas within the widest possible range of operating conditions.

The selection of maximum back-pressure is therefore <u>of essential importance and is the primary design parameter to be determined in the expander design.</u> For the operating cycle for a supercharged or turbocharged diesel, the maximum back-pressure $p_4'_{max}$ will occur with the highest air mass flow rate through the engine ($\dot{m}_{max}$) and exhaust temperature ($T_4'_{max}$), and usually coincides with rated operating conditions. The required expander volume flow capacity, which controls these parameters, can be determined from the following basic thermodynamic considerations. Assuming adiabatic expansion, the requirement of total expansion will define the volume flow rate of the expanded gases $\dot{V}_4"_{max}$, after expander as:

$$\dot{V}_{4"max} = \dot{V}_{4'(max)} \left(\frac{p_{4'max}}{p_0}\right)^{\frac{1}{\kappa}} \tag{2}$$

where:

P_0 = Ambient pressure (or muffler upstream pressure).

Since the rotational speed of the screw expander effects its overall efficiency, it also must be subject to optimizing. Both theory and practice suggest that in order to maintain satisfactory efficiency characteristics, the nominal rotational speed of the male shaft $n_{e\,max}$ should produce a rotor tip speed between 0.3 and 0.5 Mach depending on clearances used and the design of inlet and outlet ports. When $n_{e\,max}$ is determined, the expander charging volume can be calculated from:

$$V_{che} = \frac{\dot{V}_{4'max}}{n_{e\,max}} \qquad (3)$$

The total displacement can be calculated similarly:

$$V_{te} = \frac{\dot{V}_{4''max}}{n_{e\,max}} \qquad (4)$$

The geometric relation between the basic dimensions and the volume flow capacity of the screw expander is given by:

$$\dot{V} = C \quad C_{wa} \quad D^3 \left(\frac{L}{D}\right) \quad n_m \qquad (5)$$

where:

D = Rotor diameter
L = Rotor length
n_e = Male rotor, rpm.

The factors C and C_{wa} account for the influence of the rotor profile and lobe wrap angle on the displacement of the screw machine. Preliminary sizing of the expander for a given engine can be done by using Eq. (2) and Eq. (5) after assuming an $\left(\frac{L}{D}\right)$ ratio.

EVALUATION OF POTENTIAL SYSTEM PERFORMANCE

Two different computer simulation approaches have been used in predicting the theoretical performance of a screw expander compounded with a diesel engine. The first one was to develop the overall efficiency and compound power characteristics of a theoretically defined screw expander compounded with a theoretically defined adiabatic diesel operating on a high supercharged, preheat cycle developed by Cummins Diesel Co. Such a cycle is appropriate for generating after-cylinder p-V work since it provides an approximately even feeding pressure for the expander, regardless of the actual load. Figure 8 shows the simulated overall efficiency characteristics for the screw expander of the same design as shown in Figure 7 as a function of engine speed and determined for four different load settings. Figure 9 shows the expander shaft power outputs corresponding to the same conditions, while Figure 10 represents the net compounded power of the expander (after accounting for compressor power input). This analysis indicates that, except at low speeds and low loads of the system's operating range, the expander will offset the compressor parasitic losses. For most operating conditions, the screw expander shows the potential of net power contributions to the system from 10% to

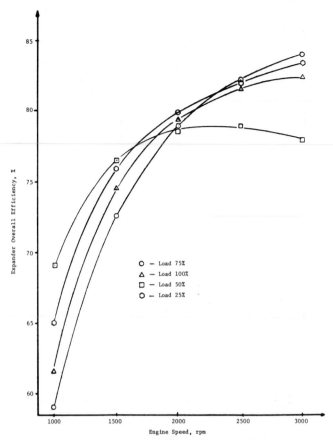

Figure 8. SCREW EXPANDER EFFICIENCY
AT VARIOUS ENGINE LOADS

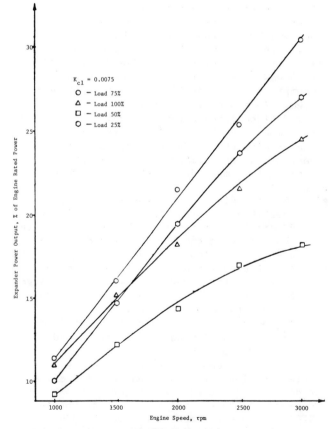

Figure 9. EXPANDER POWER OUTPUT
AT VARIOUS ENGINE LOADS

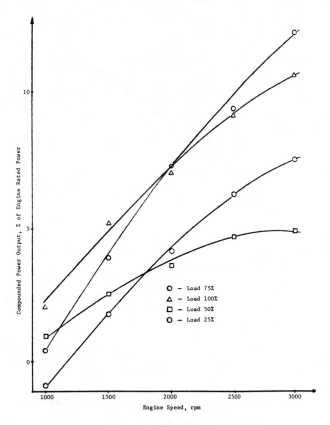

Figure 10. COMPRESSOR-EXPANDER
COMPOUNDED POWER OUTPUT

back-pressures. The general conclusion of this analysis was that the screw expander compound system has the potential to perform better than the turbocompounded baseline system by approximately 7% in terms of brake mean effective pressure and 5% in terms of brake specific fuel consumption in a wide range of operating conditions.

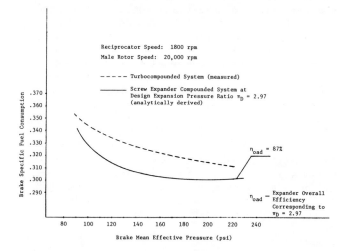

Figure 11. COMPARISON BETWEEN TURBOCOMPOUNDED
SCREW EXPANDER COMPOUNDED SYSTEM

35% of the engine brake power output. The larger relative power gains occur at low partial loads and higher speeds of the system.

Another approach in predicting the potential performance of the positive displacement system was followed in evaluating the fuel consumption in comparison to a turbocompounded system. In a simplified analysis, a theoretically defined screw expander model was combined with a model of the existing heavy-duty cooled diesel engine which has been used as a subcomponent of the experimental turbocompounded system. The simplifying assumptions for the model were:

● Positive displacement supercharger is capable of duplicating the boosting characteristics of the turbocharger used in the test hardware and has an assumed constant overall efficiency of 75%.

● Back-pressure conditions imposed on the engine by the screw expander do not change the thermal efficiency of engine but do effect its pumping losses.

● The model assumes steady-state conditions at the inlet and outlet of the expander using air as the working fluid.

This analysis showed the characteristics for several versions of screw expander with different design back-pressures. Figure 11 compares the measured brake specific fuel consumption (BSFC) of the experimental turbocompounded system and the modeled positive displacement compound system, operating at equal rated point

POTENTIAL ADVANTAGES OF POSITIVE DISPLACEMENT SUPERCHARGED COMPOUNDED SYSTEM OVER TURBOCOMPOUNDED SYSTEM

The screw-type positive displacement supercharged-compounded (PDSC) system shows several potential advantages over a turbocompounded system (TC) which, in the opinion of the authors, make this concept worthy of further investigation and development.

One of the advantages of PDSC system over TC system lies in its potentially higher overall thermal efficiency. This superiority appears to be technically feasible for the wide range of operating conditions. The basic reason for the better efficiency of PDSC systems is the higher achievable overall efficiencies of the screw compressor and expander compared to those for a turbocharger unit and a low pressure, free power turbine. Peak efficiency values for a turbocompressor and the turbine are approximately 80%, resulting in an overall efficiency for the turbocharger of 64%. The upper limit of the power turbine efficiency also is 80%. Alternatively, realistic overall peak efficiency of the screw compressor can be as high as 85% and overall peak efficiency of a screw expander can reach 90%. Some gain in thermal efficiency of the positive displacement system versus dynamic system is due to the higher friction losses inherent to high speed dynamic energy conversion process. The difference in efficiency levels would be larger for light-duty systems where the size of the turbine limits its performance. One-stage heat-to-mechanical energy conversion provided by a positive displacement screw expander also can reduce heat and gas transfer losses as compared to the two-stage turbocompounding process.

Another very attractive advantage of PDSC system is better compatibility of the rotary screw expander with the diesel engine. Achieving reasonable efficiencies in a turbine of limited size requires the turbine to run at high speeds. Practical rated speeds for free power turbines range from 50,000 rpm for larger, heavy-duty turbines to 100,000 rpm for smaller, light-duty turbines. Connecting the turbine shaft to the engine crankshaft requires a high gear ratio of about 20 to 30, which is technically viable only when complex, multi-stage power trains are used. For a screw expander regardless of system size and duty, rated speeds of the power take-off shaft can be limited to 20,000 rpm. Power transfer to the crankshaft may be accomplished by means of belt transmission, decreasing the cost of the system compared to the turbocompounded system. Overspeeding of the expander rotors, which can occur when a break in the mechanical link between the expander and engine shafts occurs, can be controlled by throttling with the expander expansion ratio modulation system. Another major advantage of a PDSC system is its potentially faster response to changes in load, due to the mechanically driven positive displacement supercharging. Elimination of turbocharger lag is also desirable to control particulate emissions.

In summary, the PDSC system represents a new approach toward highly efficient power system with increased power density, although it has no immediate counterpart in the form of proven operating hardware. It seems evident that a practical realization of the discussed advantageous features will require meeting the following conditions and technical requirements:
- Careful selection of both compressor and expander design characteristics in order to obtain optimal pressure interactions with the adiabatic diesel engine.
- Utilization of advanced profiles for the screw rotors in order to ensure high isentropic efficiency of the expander and compressor.
- The expander should produce complete expansion of the diesel exhaust regardless of system load to maximize the performance of system. Since the external expander pressure ratio is a function of the load setting, the expander should be provided with automatic internal expansion modulation.
- The screw supercharger should be provided with load modulated variable delivery characteristics to reduce the part load parasitic losses of mechanical supercharging.
- High temperature dimensional stability and structural strength of screw expander rotors as well as special requirements for high speed roller bearings can be only satisfied by the use of ceramics as structural material.

- New, inexpensive manufacturing technology for the screw rotors must be developed.

HIGHLIGHTS OF PRELIMINARY DESIGNS

As part of the preliminary evaluation of the potentials of screw machinery applied in energy recovery and supercharging service of adiabatic diesels we have assessed the design problems for such screw compressors and expanders. For that purpose we developed several sets of compressor/expander designs, using rules of screw compressor design practice with engineering considerations given by the special operating conditions imposed by the system. Our goal was to achieve an operable rather than economical design. These designs also served to define critical areas and possible simplifications well enough to identify problems other than those of common screw compressors. For compressors, the requirements are to:
- Produce low cost rotors of high precision resulting in lower than standard clearances for non-lubricated operation with anti-friction bearings.
- Resolve material problems of high pressure-ratio operation in an air-cooled design.
- Develop materials for rotors that will operate unlubricated and without synchronizing gears.

For expanders the requirements are the same, but are aggravated by very high temperature operation, the need for a proven bearing system, and composite material rotor design. The areas of desirable simplification include:
- Elimination of seals
- Use of dry bearings
- Elimination of synchronizing gears.

One specific design area which is critical to the successful development of the discussed machines is the application of advanced ceramic and composite materials and production techniques. These materials, which were characterized only by their generic thermophysical properties, have not been tried in the proposed application. However, they have been experimented with in components exposed to similar conditions, e.g., in ceramic bearings development.

Requirements of high temperature resistance (up to 900°F) and related dimensional stability for the expander made necessary the use of composite structures with both metallic and ceramic components. Ceramics are intended for the lobed shells of the rotors, housing liner, and end plates. Advanced units may also be furnished with ceramic roller bearings. Stress analysis carried out for 160 mm rotors has shown that, for the claimed strength characteristics of some experimental ceramics, structural integrity could be retained up to 20,000 rpm. Even better prospects exist for smaller rotor diameters. Table 1 summarizes the main design specifications derived for three different base engines. A longitudinal view of the expander for the HD 400 hp engine is shown in Figure 12.

Table 1. DESIGN SPECIFICATIONS FOR VARIOUS SCREW EXPANDERS

Type of Base Engine	Rated Performance, hp/rpm	Rated Gross Power of Expander, hp/male rotor rpm	Main Dimensions				
			Rotor Diameter, mm (inches)	Rotor Length, mm (inches)	Overall Length, mm (inches)	Overall Height, mm (inches)	Overall Width, mm (inches)
LD Adiabatic Diesel	92/3000	30/20,000	98 (3.858)	98 (3.858)	340 (13.385)	180 (7.086)	230 (9.055)
HD Cooled Diesel	400/1900	100/20,000	160 (6.299)	240 (9.448)	540 (21.259)	280 (11.023)	360 (14.173)
HD Adiabatic Diesel	750/3500	200/18,000	204 (8.0315)	306 (12.047)	800 (31.496)	350 (13.779)	460 (18.116)

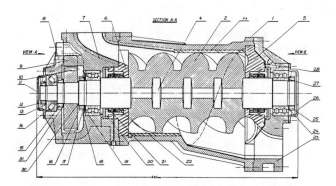

Figure 12. LONGITUDINAL VIEW OF THE EXPANDER FOR THE HD 400 ENGINE

SUMMARY OF RESULTS AND CONCLUSIONS

Our theoretical system analysis and preliminary designs carried out for several sizes of positive displacement supercharged/compounded systems have proven the technical feasibility and attractive performance potentials of applying screw machinery. This is based on the following results of our evaluation:

1. <u>Performance</u>. Positive displacement supercharging and compounding is especially compatible with a light-duty, high BMEP adiabatic diesel. In this configuration, a screw supercharger, screw expander, and diesel reciprocator can be compounded into a low-cost, high power-density power plant with excellent fuel economy and performance characteristics. The better overall efficiency of small positive displacement screw expanders over small turbo-expanders, the lower operating speeds, and elimination of turbocharger lag are the key reasons for promoting the concept of light-duty, positive displacement supercharged/compounded power systems. Even for larger systems, positive displacement machinery holds an efficiency advantage of 5% to 10%, superior expander-reciprocator speed compatibility, and dynamic response characteristics. To assure the efficiency advantage of such systems the positive displacement supercharger should be designed with a load-controlled charge modulation system to alleviate the problem of part-load operation parasitic losses.

2. <u>Design and Engineering</u>. The key to the successful application of screw machines for supercharging and compounding, particularly adiabatic diesels, lies in the ability to develop new design approaches using high-temperature-resistant materials with low thermal expansion and friction factors to survive adverse operating conditions and to allow lower production costs than currently encountered. It is likely that experience from current material and component development in other fields can be successfully used. The preliminary structural analysis has shown that silicon nitride ceramics under development offer promising mechanical properties and may be used in the design of composite rotors and other less critical components. Significant progress achieved in high speed, ceramic ball and roller bearings offer the possibility of solving the problem of unlubricated shaft bearings.

Based on these results, we have concluded that:

- Particularly for the light-duty prime-movers (with high power output density), for which simplicity, compactness, and low cost compliment the essential requirement of high performance effectiveness over a wide range of operating conditions, the better overall thermal efficiency and compatibility of the positive displacement expander/diesel engine system make the double-screw expander a very attractive candidate.

- The positive displacement expander/diesel engine compounded system requires a high level of supercharging. The mechanically-driven screw compressor has the potential of being a compact and efficiency supercharger, if the problem of part load losses is solved.

840432

Post Densified Cr_2O_3 Coatings for Adiabatic Engine

Jeffrey Carr and Jack Jones
Kaman Sciences Corp.

ABSTRACT

A Cr_2O_3 impregnation treatment process can be used to densify and strengthen existing coatings and materials and to produce new coatings for use in adiabatic engines.

Plasma sprayed zirconia densified with Cr_2O_3 exhibits an 87% lower wear rate than an undensified plasma sprayed zirconia coating, and a thermal conductivity value of 9.67 BTU - in/hr-ft^2-°F at 1832 °F. A Cr_2O_3 densified ceramic coating exhibits a thermal conductivity value of 14.9 BTU - in/hr-ft^2-°F at 1500 °F.

THE Cr_2O_3 IMPREGNATION TREATMENT and coating processes consist of a family of effective techniques centered around a unique principle of oxygen bonding for use with most metals and ceramic substrates. The treatments are done at relatively low temperatures but nevertheless result in high corrosive attack resistance, high wear resistance, high hardness, and can be used in many applications requiring temperatures much higher than the original cure temperature. In general materials and substrates treatable with this process include many solid ceramic materials, metals, plasma and flame spray coatings. Examples of ceramics include refractory oxides, carbides and silicides. Metals include powdered metals, solid metals, both high and low carbon steels, 400/300 series stainless, titanium, Inconel, Monel, brass, and beryllium copper. Other porous materials and coatings, such as hard chrome plate, flame sprayed and plasma sprayed coatings are also treatable with this process. In each case where a supplementary coating is treated with the Cr_2O_3 process hardness is increased, bond strengths are also generally increased along with the corrosion resistance and wear characteristics. Also of importance is the wear, which is reduced in the cases of powdered metals, flame sprayed and plasma sprayed coatings which were treated with the ceramic process. The resultant treated products also exhibit high thermal shock resistance. The density can be varied based on the process changes and composition changes that are available with this treatment. This results in a ceramic treating process which can be tailored to fit each specific application and end use. Some applications require higher thermal expansion, some require lower, some require non-wetting conditions. All of these can be built into this coating process. A more detailed discussion of each specific process, including processing techniques, applications, end uses, and characteristics of the product will follow.

The following Cr_2O_3 impregnation process as used in diesel/automotive engines is performed to densify and strengthen a porous base material, either coatings or solid ceramic material.

The chrome is applied in liquid form made up of hexavalent chromium and water. The liquid is then applied in a variety of ways: either spraying, painting or immersion.

Capillary action will pull the chrome liquid into the open pores, thereby filling the pores with the liquid chrome. The coating is then fired at a specific temperature (1000°F), which drives off the water and converts the hexavalent chrome into Cr_2O_3 at the same time generating an oxide bond. Repeated applications will increase the amount of chrome deposited within the body until such time as all the open porosity is filled or sealed off.

The first application of the Cr_2O_3 impregnation system is in the area of porous ceramic coatings such as plasma flame sprayed coatings, slurry applied coatings, and hard chrome plate. Potential areas of use include cylinder liners, pistons, rings, heads, and exhaust ports/manifolds.

Use of the Cr_2O_3 process will improve specific properties of the above coatings. In the case of plasma sprayed zirconia an average 87% decrease in wear rate is exhibited. The plasma sprayed zirconia (Y_2O_3 stabilized) was run on a Falex 6 wear tester against itself as was the Cr_2O_3 impregnated plasma sprayed zirconia. (See Figures 1 and 2.)

It is interesting to note that as the pressure increases on the Cr_2O_3 impregnated plasma sprayed zirconia the wear rate was reduced. One theory for this is that the chrome removed is broken down and acts as a solid lubricant.

Thermal conductivity figures on this same type coatings (.025" thick) noted an increase in the conductivity figures after impregnation. There is a 43% increase at 1832°F increasing from 6.78 → 9.67 BTU - in/hr-ft^2-°F. This is due to the fact that the pores are now filled with Cr_2O_3 instead of air. (See Table 1 and Figure 3.)

Another modification of this system is the sealing of pores which prevents infusion of fuels into the coating which could ignite prematurely. This sealing will also prevent corrosives from penetrating the coating and attacking the substrate, which leads to coating failure. (See Figure 4.)

Use of the application described include cylinder liners and piston caps and valves. The coating is sketched in Figure 5.

The Cr_2O_3 impregnation system is also used with slurry applied coatings. Kaman's SCA-1000 coatings, for example, derive its wear resistance and corrosion resistance from this process.

Briefly, the SCA system utilizes silica, alumina and chrome which are blended to form a composite with roughly 50-60% porosity. The Cr_2O_3 is used to fill the porosity and harden the coating. (See Figure 6.) The presence of Cr_2O_3 will help to build at oxide bond. During the firing a diffusion takes place between the chrome and the substrate iron. (See Figure 7.) This ferric chromate bond layer has been measured on epoxy pull tests to be over 10,000 psi which is the limit of the test. Photos of an SCA coating are in Figures 8 and 9.

Much the same as with plasma coatings, the Cr_2O_3 impregnation system seals off the porosity and prevents corrosives from attacking the substrate. (See Table 2.)

Use of this coating has been in application requiring high wear resistance and high heat resistance such as cylinder liners and valve stems and guides.

A modification to this system will increase the thickness from maximum .005" to a range of .020-.100". Again, the Cr_2O_3 impregnation system is used to seal the outer porosity. This coating is designed for use as an alternative to current coating systems which cannot be applied to intricately shaped I.D. surfaces such as the exhaust ports inside a cylinder head or an exhaust manifold. This thick coating utilizes

the inherent porosity to improve its K factor. Adjustments to the density of the coating (i.e. further densification) will result in the lowering of the conductivity results. (See Table 3 and Figure 10.)

The final area to be discussed that this process can be used in is the area of solid ceramics. Similar results can be achieved as is seen with plasma coatings: that is, improvement in wear and corrosion resistance. The case examined compares a 99% Al_2O_3 body both as fired and as densified. Tests were run on a standard rub shoe test. (See Figure 11.)

In the case where a dense surface is desired while still retaining porosity inside, the Cr_2O_3 impregnation method can be adjusted so only the surface is treated. Conversely, utilizing pressure or vaccuum impregnation techniques can force the Cr_2O_3 throughout the body.

TABLE 1

THERMAL CONDUCTIVITY CALCULATIONS

Sample (No.)	Temp. (°C)	Density (gm cm^{-3})	Specific Heat (W s gm^{-1}K^{-1})	Diffusivity (cm^2 sec^{-1})	Conductivity (W cm^{-1}K^{-1})	Conductivity (BTU in hr^{-1}ft^{-2}F^{-1})	Temp. (°F)
1	23	5.412	0.475	0.00550	0.0141	9.80	73
	100	5.412	0.518	0.00519	0.0145	10.09	212
Zr O$_2$	200	5.412	0.558	0.00497	0.0150	10.41	392
+	300	5.412	0.584	0.00479	0.0151	10.56	572
Cr$_2$O$_3$	400	5.412	0.602	0.00461	0.0150	10.41	752
	500	5.412	0.614	0.00443	0.0147	10.21	932
	600	5.412	0.630	0.00429	0.0146	9.91	1112
	700	5.412	0.632	0.00418	0.0143	9.78	1292
	800	5.412	0.634	0.00411	0.0141	9.69	1472
	900	5.412	0.636	0.00406	0.0140	9.69	1652
	1000	5.412	0.638	0.00404	0.0139	9.67	1832
2	23	5.320	0.470	0.00520	0.0130	9.01	73
Zr O$_2$	100	5.320	0.514	0.00478	0.0131	9.06	212
	200	5.320	0.555	0.00420	0.0124	8.60	392
	300	5.320	0.578	0.00370	0.0114	7.89	572
	400	5.320	0.594	0.00324	0.0102	7.10	752
	500	5.320	0.614	0.00302	0.00986	6.84	932
	600	5.320	0.626	0.00288	0.00959	6.65	1112
	700	5.320	0.627	0.00283	0.00944	6.55	1292
	800	5.320	0.630	0.00281	0.00942	6.53	1472
	900	5.320	0.632	0.00283	0.00952	6.60	1652
	1000	5.320	0.634	0.00290	0.00978	6.78	1832

Table 2 Chemical Durability Test

Chemical	Volume % Concentration	Time of Test
HCL	100%	3 hours
HF	100%	1 hour
HF	10%	72 hours +
H_2SO_4	17%	72 hours +
HCL	17%	72 hours +
HF	10%	72 hours +
H_3PO_4	17%	72 hours +

Table 3   Thermal Conductivity Calculations

Temp. (°C)	Density (gm cm^{-3})	Specific Heat (W S gm^{-1}K^{-1})	Diffusivity (cm^2 sec^{-1})	Conductivity (W cm^{-1}K^{-1})	Conductivity (BTU in hr^{-1} ft^{-2}F^{-1})	Temp. (°F)
23	2.933	(0.735)	0.0177	0.0382	26.5	73
93	2.933	0.776	0.0140	0.0319	22.1	200
204	2.933	0.845	0.0114	0.0283	19.6	400
315	2.933	0.903	0.0098	0.0260	18.0	600
427	2.933	0.938	0.0086	0.0237	16.4	800
538	2.933	0.955	0.0077	0.0216	15.0	1000
649	2.933	0.964	0.0075	0.0212	14.7	1200
760	2.933	(0.973)	0.0075	0.0214	14.8	1400
815	2.933	(0.977)	0.0075	0.0215	14.9	1500

( ) = extrapolated

UPPER WASHER MATERIAL: PLASMA TECHNICS PT013 COATING ON 1018 STEEL
LOWER DISC MATERIAL: PLASMA TECHNICS PT013 COATING ON 1018 STEEL
TEST: FALEX-6 OSCILLITORY TEST AT 230 cpm, 90°
CONTACT PRESSURE: 25 psi, 50 psi, & 100 psi
TEMPERATURE: ROOM TEMP., 25° C
OTHER: UNLUBRICATED; DATA AT 11 MIN. ELAPSED TIME

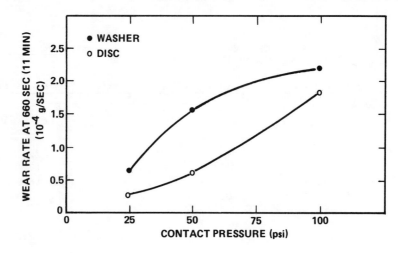

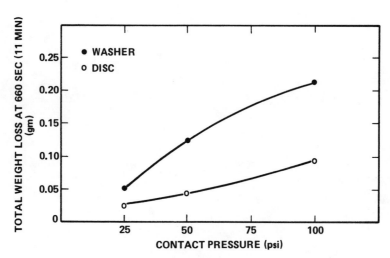

FIGURE 1

WEAR TEST

UPPER WASHER MATERIAL:	Cr_2O_3 IMPREGNATED PLASMA TECHNICS CTBC
LOWER DISC MATERIAL:	Cr_2O_3 IMPREGNATED PLASMA TECHNICS CTBC
TEST:	FALEX-6 OSCILLITORY TESTING AT 230 cpm, 90°
CONTACT PRESSURE:	25 psi, 50 psi, & 100 psi
TEMPERATURE:	ROOM TEMP., 25° C
OTHER:	UNLUBRICATED; DATA AT 33 MIN. ELAPSED TIME

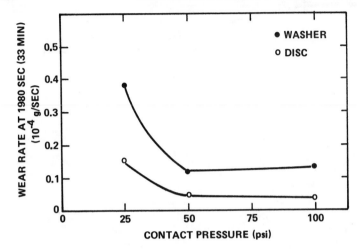

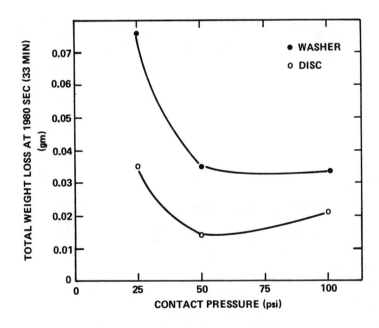

FIGURE 2

WEAR TEST

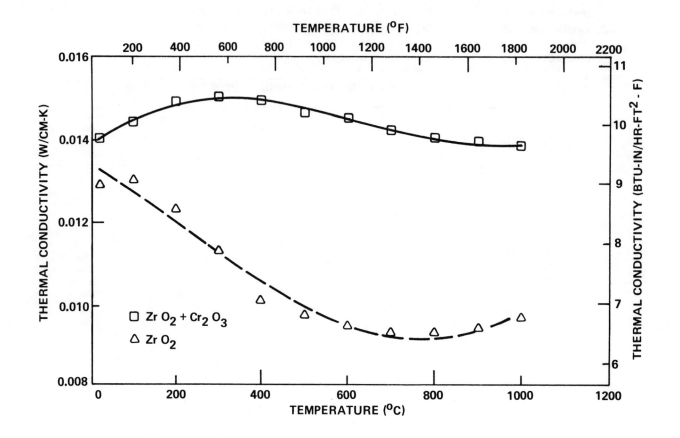

FIGURE 3

THERMAL CONDUCTIVITY OF Zr O_2

WITH AND WITHOUT Cr_2O_3

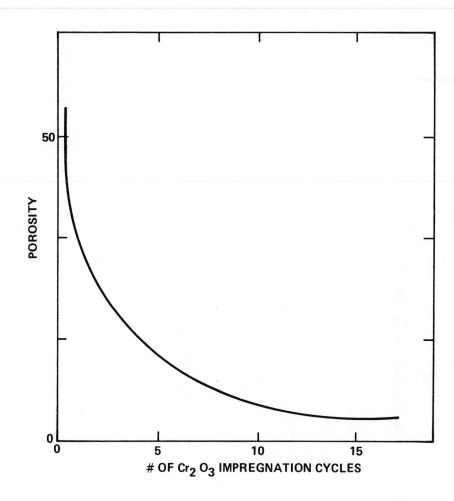

FIGURE 4

POROSITY

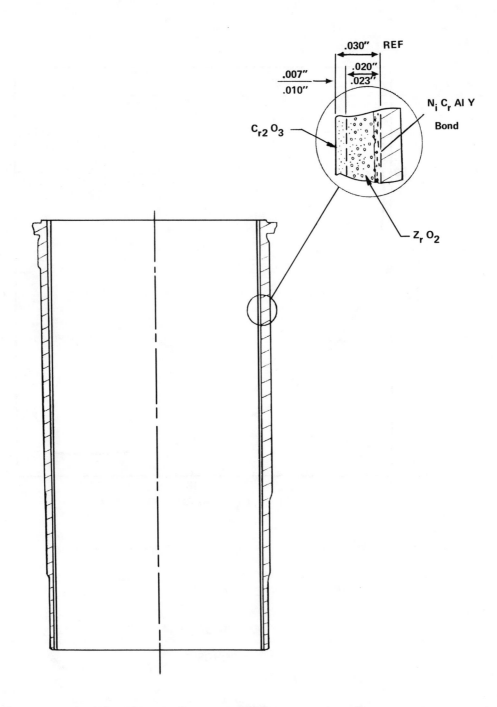

FIGURE 5

COMPOSITE ZrO$_2$ – Cr$_2$O$_3$ COATED CAST IRON LINER

(NOT TO SCALE)

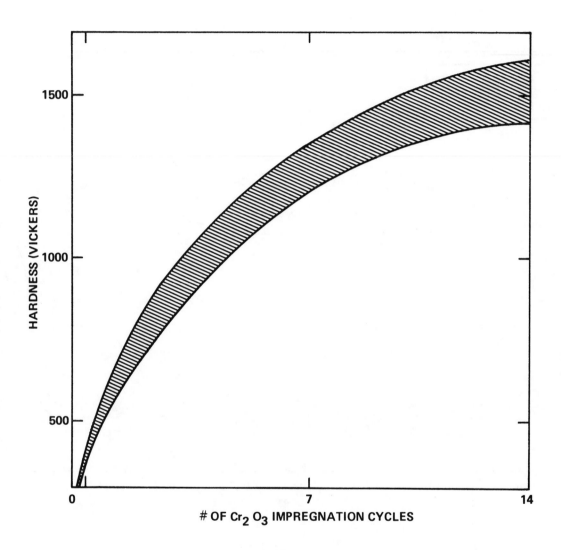

FIGURE 6

HARDNESS

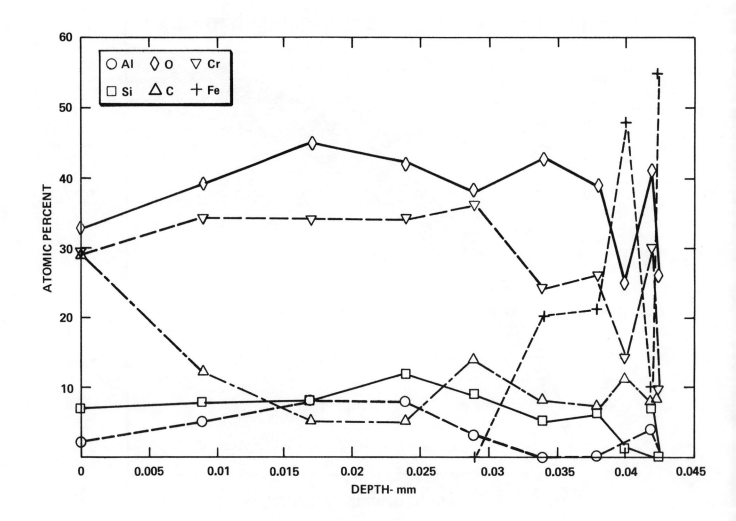

FIGURE 7

KAMAN SCIENCES - CERAMIC COATED ROD

FIGURE 8

METAL/COATING INTERFACE
100X

FIGURE 9

TOP OF COATING
100X

67

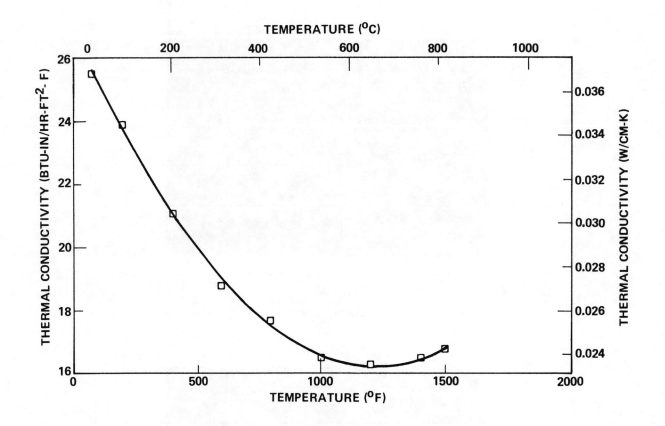

FIGURE 10

THERMAL CONDUCTIVITY

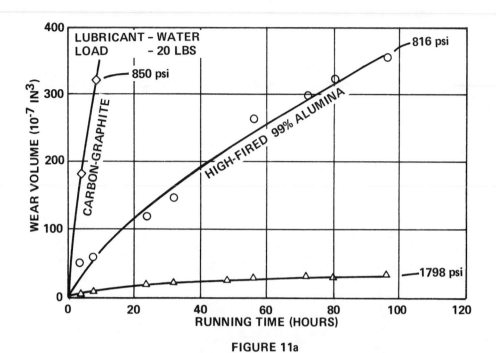

FIGURE 11a

RUB-SHOE WEAR COMPARISONS OF CARBON-GRAPHITE,
HIGH-FIRED ALUMINA AND K-RAMIC DENSIFIED 99% ALUMINA

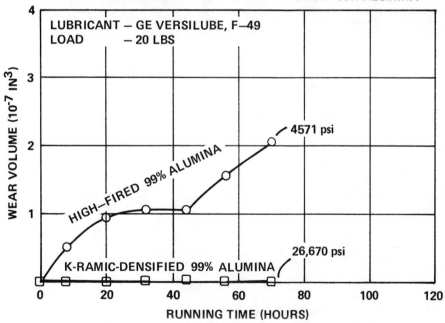

FIGURE 11b

RUB-SHOE WEAR COMPARISONS OF HIGH-FIRED ALUMINA
AND K-RAMIC-DENSIFIED 99% ALUMINA

840433

Engine Rig for Screening Ceramic Materials

S. G. Timoney
University College Dublin (Ireland)

ABSTRACT

Ceramic materials used either as solid components or as coatings on metal components provide a bewildering choice of property characteristics to the designer. Application in Diesel engines to permit efficient, high temperature, no-coolant operation is seriously hampered by lack of test experience with these materials and lack of a comprehensive data base on performance characteristics and properties in relation to engine conditions. The paper describes an engine designed specifically for testing ceramic materials in piston and liner applications, and the advantages of this approach in practice.

ENGINEERING CERAMICS for use in Diesel engines are probably the most exciting materials under development today. Japan plans to spend at least $60 million, the U.S.A. $80 million, and the European Economic Community $40 million over the next 3 to 5 years, on this business opportunity said to be worth $15 billion/year, (1,2,3)*. Much of the interest derives from the work initiated by Kamo (4) in 1975, when he suggested the possibility of substantial improvements in the energy conversion effectiveness of Diesel engines through the adoption of what he called the "Adiabatic Turbo-compound Diesel".

Inspired by Kamo's initiative in the U.S.A., work on a "No-Coolant Diesel Engine" was begun at University College Dublin (5,6) under contract to the EEC

*Numbers in parentheses designate references at end of paper.

in 1976. This work has now led to the design of a specific research engine for the development and evaluation of ceramic materials for use in so-called "Adiabatic" or "No-Coolant" Diesel engines.

NO-COOLANT ENGINES

Engines designed to run without any conventional cooling system must be able to withstand temperature in the order of 800°C on the surfaces of the combustion chamber. Conventional engine materials will not sustain these conditions. Ceramic materials, on the other hand, are suited to these high temperature conditions, but not to the mechanical loading, nor indeed in some ceramics to the thermal shock loading, that can often exist in an engine combustion chamber. A very real difficulty exists in the scarcity of experience and knowledge in regard to the engineering properties of ceramic materials. Diesel engine applications, whether in solid ceramic form or as protective coating on metals, subject these materials to extremely complex conditions. During each cycle of the engine the combustion chamber environment changes from relatively cool air to very high temperature combustion products containing complex mixtures of gases such as CO_2, CO, N_2, O_2, H_2O, NO_x, C_nH_m, etc. At the same time the pressure varies rapidly between 1 and 200 atm. and the gas temperature between 25 and 2500°C, subjecting the chamber wall to substantial fluctuations in stress pattern.

Unfortunately, the conditions in each particular type of engine are more or less unique and the stresses set up depend very much on the specific shape of the combustion chamber walls including the piston crown. Consequently, test results for a ceramic material in one particular engine may not, in fact, be directly comparable with those for another material, or indeed for the same material, in another design of engine.

Most of the tests in the U.S.A. to date have been carried out with a Cummins NHC 250 engine of 14 litre capacity (4). In Japan, NKG has carried out work on a small 50cc two-stroke engine. Some other work in Japan has been reported by Kyoto on a modification to a Deutz three cylinder engine of 2.7 litre capacity, (1). None of these engines have remotely similar combustion chambers and comparison of results is therefore of doubtful value.

MATERIALS

Ceramic materials, such as SiC and Si_3N_4 have very varying characteristics as regards strength, density, rubbing friction and other properties. The precise values of these parameters are greatly effected by the forming process, the heat treatment, the surface finish and so forth; the behaviour in an engine is also influenced by the shape of the part, the thermal and load stresses applied to it and the manner in which they are applied. Little is known about the effect on the properties of ceramics of the hot and high pressure combustion gases, during the operating cycle.

Clearly a lot of information can, and must be, acquired by the testing of standard specimens in the usual manner for the normal physical properties. But, the difficulty, however, lies in relating these properties to the behaviour of the ceramic materials when formed into parts for an engine. Unlike metals, ceramics are very inflexible materials, and when they are over-stressed they rupture without detectable elongation by comparison with the behaviour of even the most brittle metals.

Furthermore, failure generally results in a multiplicity of fractures and an extensive shattering that makes failure analysis very difficult. The fragments of a ceramic liner after failure are shown in fig. 1, and illustrate this point clearly.

Serious consideration has been given to the design of test rigs which would include high temperature cycling in an environment of combustion products and even some force inputs such as might be encountered in engine parts. The rig concepts become very complex and so close to an engine that a decision was made at University College Dublin to design an engine that would substitute effectively for such a rig. This engine would have to use low cost shapes, capable of being formed from a wide variety of ceramic materials and typical of what would be useful in conventional engine design practice.

FIG. 1 FRAGMENTS OF A CERAMIC PISTON AND LINER FAILURE

STANDARDISED TESTING

In the 1920's similar difficulty was experienced in evaluating the detonation response of different fuels for petrol and Diesel engines. The characteristics of fuels determined by the conventional laboratory parameters and procedures did not distinguish the good from the bad in relation to detonation in actual engines. The interested bodies in the petroleum and automotive industries set up a committee to examine the problem and the Co-operative Fuel Research Engine (CFR engine) was conceived. These engines, and the test procedures established for their use, are still, 60 years later, the work horses of the petroleum industry in regard to Octane and Cetane ratings of petroleum fuels.

It seems that the time has now come to have a similar standard engine for evaluation of different ceramic materials under the conditions of use in a Diesel engine. Such a standard engine, if used by ceramic manufacturers and by engine manufacturers, could lead to a rapid build-up of information on the behaviour of ceramics under engine combustion chamber environmental conditions. Different types of ceramics, subjected to different manufacturing processes and heat treatment, would be formed into similar parts and tested under similar controlled conditions. The materials would, of course, be evaluated for the conventional standard physical properties also and the accumulation of results for their subsequent behaviour in the engine tests would eventually begin to shed some scientific light on the design choices that effected behaviour. Comparable results could be obtained for widely differing ceramic materials used, either as solid forms or as protective coatings on metallic forms, in an engine environment.

In order to attain maximum flexibility in the choice of material and production process it is desirable that the shapes into which the parts have to be made are relatively simple. Clearly, a complex shape such as one might encounter with the combustion chamber of a conventional indirect injection engine could give rise to complex and unique stress raisers. Such unqualified inputs would greatly increase the difficulty of carrying out truly comparative tests as between one ceramic material and another, and would be unacceptable.

The necessity for metallic valves in a conventional four stroke cycle engine layout also introduces an unnecessary complication by comparison with the use of a ported two-stroke cycle engine arrangement for filling and emptying the cylinder. Furthermore, the problem of clamping the cylinder head to the cylinder block, if these are to be made in ceramic materials, or even have substantial ceramic components in them, is made very difficult by the difference in coefficient of thermal expansion between the different ceramic materials and metals, and should, if possible, be avoided.

TEST ENGINE

The opposed piston two stroke cycle cylinder layout has much to offer as a concept for a standardised engine for evaluation of ceramic material behaviour in a Diesel engine environment. Early work under a contract from the Commission of the European Economic Community (EEC), (5,6) led to the design and development of the small (500cc) single cylinder opposed piston engine shown in fig. 2. This design was used to test cylinder and piston assemblies in various ceramic materials α SiC, Si_3N_4, and β Spodumene, Lithium Alumino Silicate (LAS). The α SiC

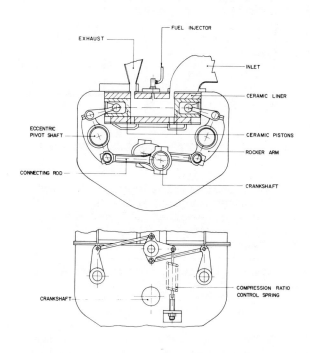

FIG. 2 500cc RESEARCH ENGINE

was the only material with which substantial running time was achieved. The
satisfactory running of an uncooled,
unlubricated and ringless SiC piston in
a SiC cylinder liner early in 1982, has
demonstrated some of the attractive
possibilities that ceramic materials
offer to the Diesel engine designer.
The α SiC piston, running unlubricated
in the liner, for instance, was dis-
covered with great interest to have a
significantly lower friction drag than
that of a conventional, lubricated piston
in a cast-iron liner, fig. 3 (7).

Δ– LUBRICATED METAL ENGINE

Z– UNLUBRICATED SiC ENGINE, INITIAL FIRING

O– UNLUBRICATED SiC ENGINE, REDUCED PISTON
    LINER CLEARANCE

X– METAL ENGINE WITH PISTONS DISCONNECTED

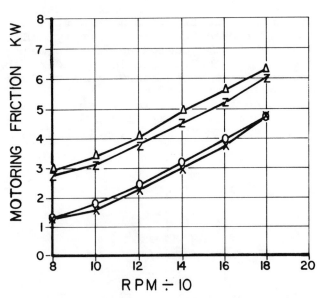

FIG. 3 FRICTION LOSSES COMPARED

The first engine design did not
itself address the requirements of a
standard test engine for ceramic
component evaluation, but it did
demonstrate the desirability of
having such an engine and it did
indicate some of the special features
that such an engine should have.

STANDARD TEST ENGINE

A new engine of 1 litre capacity
per cylinder has now been designed
specifically as a laboratory engine
for the evaluation of solid ceramic
parts or ceramic coated metal parts
in a Diesel engine environment.

The advantages offered by this design
are:-

1. The engine cylinder containing
two pistons and the injection nozzle may
be removed from the engine without
disturbing any bearings in the running
gear of the engine; this ensures that
tests give a direct comparison of
cylinder - piston behavioural changes
and are not influenced by running gear
changes of any type.

2. The cylinder liner is of a form
suited to manufacture in a great variety
of ceramics and by a variety of production
processes. The pistons, too can be made
in different ways and with different
materials, fig. 4. Both component
concepts lend themselves to manufacture
as metals suited to coating with protective ceramic surfaces, either as
insulations and as rubbing surfaces,
or as both.

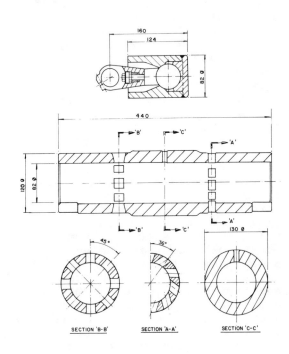

FIG. 4 CERAMIC PISTON AND LINER

3. The design for the engine
incorporates a variable compression
ratio mechanism which may be locked at
any desired fixed value or allowed to
vary automatically in response to peak
cycle pressure. This variation is
controlled by a spring and the result
is tantamount to provision of an
elasticity in the combustion chamber
walls which would dissipate those
sharp shock loads which often derive
from combustion, (8).

4. Provision is made for variation of the injection timing through 10°/15° crank angle and this permits the carrying out of experiments to evaluate the advantages of a very hot combustion chamber, in reducing the combustion delay time.

The layout of the engine is shown in fig. 5.

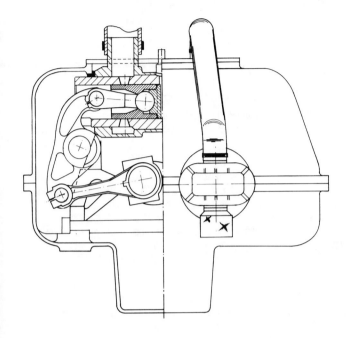

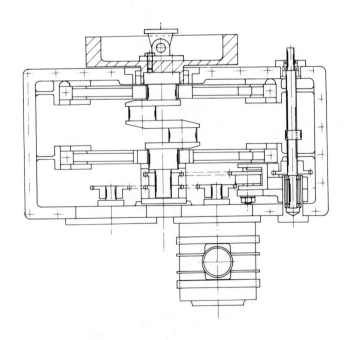

FIG. 5 ENGINE LAYOUT

BULK DENSITY	Kg/m^3
BENDING STRENGTH (4 Point)	MP_a
TENSILE STRENGTH	MP_a
IMPACT STRENGTH	KJ/m^2
COMPRESSIVE STRENGTH	MP_a
WEIBULL MODULUS	-
YOUNGS MODULUS	MP_a
POISSONS RATIO	-
HARDNESS HVIO	N/mm^2
FRACTURE TOUGHNESS	$MN/m^{3/2}$
FRACTURE IMPACT ENERGY	J
OPEN POROSITY	%
AVERAGE PORE SIZE	μ m
SURFACE ROUGHNESS (R_z)	μ m
(R_a)	μ m
COEFFICIENT OF FRICTION	-
THERMAL EXPANSION	$10^{-6}/K$
THERMAL CONDUCTIVITY	W/mK
SPECIFIC HEAT	J/kgK
TEMPERATURE LIMIT	K

THERMAL STRESS RESISTANCE FACTORS:

$$R_1 = \frac{\text{Tensile Strength}}{\text{Youngs Modulus x Thermal Exp.}} \quad K$$

$$R_2 = R_1 \text{ x Thermal Conductivity} \quad W/m$$

THERMAL SHOCK RESISTANCE
(Cold Water Quench) Δ T

FIG. 6 PHYSICAL PROPERTIES
OF CERAMIC MATERIALS

PHYSICAL CHARACTERISATION

Tests are currently carried out to evaluate the property characteristics of ceramic materials, as listed in fig. 6.

There is little or no data or experience which relates the response of materials under engine operating conditions to these standard test criteria and consequently, useful prediction of behaviour cannot be assumed at the design stage. Mechanical failure of ceramic components will usually derive from a defect. Defects, such as pretest flaws, are introduced during manufacture, while others develop during exposure to aggressive environments, such as oxidants, projectile impact, etc. It can be shown approximately that if a theoretically ideal ceramic has a tensile strength of 900 MPa, then were it to have pores 1 μm in diameter its strength would reduce to 750 MPa. Pore sizes of 10 μm would reduce this to 500 MPa, and of 100 μm to 250 MPa, as shown in fig. 7. The pore size is thought to be critical yet it is very difficult with equipment available today to measure pore size on a quantity production basis.

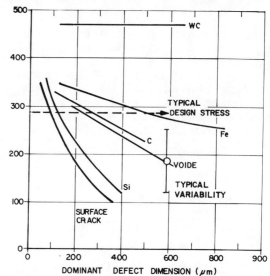

FIG. 8 DEFECT EFFECTS ON FRACTURE STRENGTH

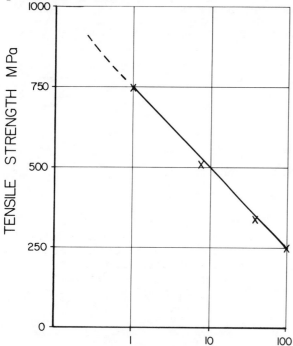

FIG. 7 AVERAGE PORE SIZE ($\log_n$) μm

Furthermore, it is not known whether the reduction in tensile strength as a function of pore size, as shown in fig. 7, will, in fact, be reflected in engine operation. Little is known, for instance, about the effect of pore size as distinct from density of pores, or number of pores per unit of specific volume. Excellent analytical and experimental work on the structural reliability of ceramic materials has been reported by Evans (8) and is summarised in fig. 8. The surface crack is seen here to be much more destructive than a void or spherical inclusions of Fe, Si or C. At the typical design stress a surface crack of 50 μm is as serious as a void of 200 μm or an iron inclusion of over 400 μm. Impurities, such as might derive from process additives, do clearly

effect the fatigue life of ceramics, and it is possible that impurities deriving from the products of combustion at high temperature in an engine might have similar undesirable effects. Only testing in engines can give acceptable answers to these questions.

Even less is known about the characteristic behaviour of ceramic coatings on metal pistons and liners. It is assumed that the compatability of coefficients of expansion between the metal and the coating is of paramount importance. Far too little is known, however, about the possibilities that exist for varying such characteristics across the thickness of a ceramic coating. No systematic analysis has been made of coating characteristics in relation to their behaviour in an engine environment.

Little is known about the effect of surface finish on the behaviour of ceramic interfaces which are rubbing together, nor, indeed, about its effect on failure sensitivity. While these factors could be evaluated outside an engine environment, it is possible that gas film lubrication may be greatly effected by surface finish also. Furthermore, the surface finish may alter the effect of the combustion gases on the ceramic materials.

MANUFACTURING PROCESSES

One of the problems for the engine designer who wishes to use ceramics is the very great choice, not only of material composition itself, but also of manufacturing process. Ceramics can be made in 10 or more processes and the choice of process will depend on the shape of product required and the material characteristics required.

76

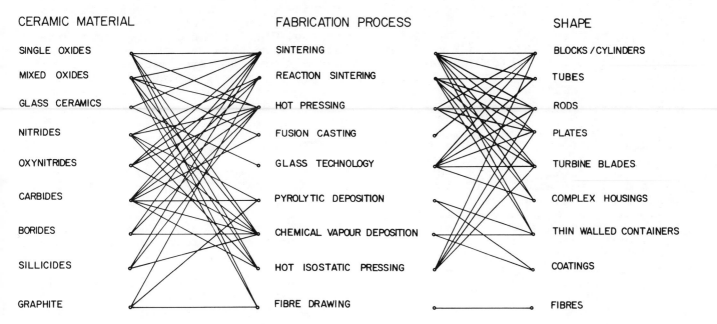

CERAMIC MATERIAL

SINGLE OXIDES
MIXED OXIDES
GLASS CERAMICS
NITRIDES
OXYNITRIDES
CARBIDES
BORIDES
SILLICIDES
GRAPHITE

FABRICATION PROCESS

SINTERING
REACTION SINTERING
HOT PRESSING
FUSION CASTING
GLASS TECHNOLOGY
PYROLYTIC DEPOSITION
CHEMICAL VAPOUR DEPOSITION
HOT ISOSTATIC PRESSING
FIBRE DRAWING

SHAPE

BLOCKS/CYLINDERS
TUBES
RODS
PLATES
TURBINE BLADES
COMPLEX HOUSINGS
THIN WALLED CONTAINERS
COATINGS
FIBRES

FIG. 9 AVAILABILITY OF SHAPES OF CERAMIC MATERIAL

Some processes are suited to small batch production while others are limited to long runs, where heavy tooling costs can be spread over a large number of products. Some of the possible combinations are shown in fig. 9, (9). Furthermore, if one examines any one particular ceramic, such as reaction sintered silicon nitride, this material can be manufactured into products by dry pressing, isostatic pressing, slip casting, extruding or injection moulding. Each process has its own particular advantage. The finished product, however, will have greatly varying properties which depend not only on the forming process but also on the subsequent firing process.

Similar product shapes manufactured by different processes will have to be subjected to directly comparable test programmes in an engine environment and that can easily be accommodated in the proposed test engine.

CONCLUSION

Although pockets of work have been in hand over a period of nearly 10 years the results to date in regard to a build-up of scientific knowledge of ceramic materials for engine applications is very disappointing. This paper is proposing a standardisation of material

evaluation testing by the provision of a generally accepted and unique engine for such testing. Engine tests, however, are only one aspect of the required programme. Much basic work is required in high temperature chemistry to allow better control of uniformity and re-producibility in pre-specified material structures. Success in this work would provide materials with strictly controlled diffusion of ingredients and with tightly limited size and volumetric distribution of pores.

Production process research is required to minimise cost and quality assurance and inspection techniques must be developed to match the serious responsibility of automotive component supply. One cannot but shiver at the thought of millions of engines on call-back for catastrophic failure of ceramic components after 50,000 km in service - and, in our present knowledge of these materials, this could happen to the intrepid pioneer of a ceramic production engine!

Actual running hours in the engine environment are of paramount importance in evaluating service behaviour and in the correlation of laboratory specimen measurements with fracture, abrasion, wear, etc., in the engine test. Care-fully controlled tests are needed on a multitude of basic but, as yet, unknown design parameters. Can all ceramics be run face to face without lubrication ? What clearance is required with different ceramics and in different size and bulk configurations ? Will the hot products of combustion change the structure of contact ceramic surfaces ? What fatigue behaviour can one expect from thermal, or force cycling, or from a combination of both ? And there are many more similar and, as yet, un-answered questions. The problems created in the development of a new product in a metal are minimised by the material variations that can be achieved by limited alloying and heat treatment. In ceramics a bewildering variety of materials and composites are on offer, each of them amenable to wide variations in properties with relatively small changes in raw material production process or design configuration.

A resume of the probable research requirements along the odyssey from raw powder to production engine is depicted in fig. 10. If the end of this journey is to be reached without excessive expenditure of scarce resources in people and money, a planned approach must be put forward to avoid duplication, and indeed waste, of effort.

REFERENCES

1. J. Hartley, "Japan Gears Up for Ceramics" The Engineer (London), 29 April 1983.

2. "Advanced Materials Research for Transport Applications" EEC Document, March 1983.

3. J.W. Dizard, "The Amazing Ceramic Engine Draws Closer" Fortune International, 25 July 1983.

4. W. Bryzik, and R. Kamo, "TACOM/ Cummins Adiabatic Engine Program" SAE Paper, 830314, Detroit, 1983.

5. S.G. Timoney, "Preliminary Experiences with Ceramic Pistons and Liners in a Diesel Engine" Proc.Int. Energy Agency Conference, Berlin, April 1981, pp.2411-2420.

6. Ed. S.G. Timoney, "Survey of the Technological Requirements for High Temperature Materials R & D : Diesel Engines" EEC Monograph No. EUR 7660 EN, 1981.

7. S.G. Timoney and G. Flynn, "A Low Friction, Unlubricated SiC Diesel Engine", SAE Paper 830313, Detroit, 1983.

8. A.G. Evans, "Structural Reliability: A Processing-Dependent Phenomenon" Jnl. of the American Ceramic Society, V.65, No.3, March 1982.

9. S.G. Timoney, "No Coolant Diesel Engine" EEC Report No. EUR 8358 EN, 1983.

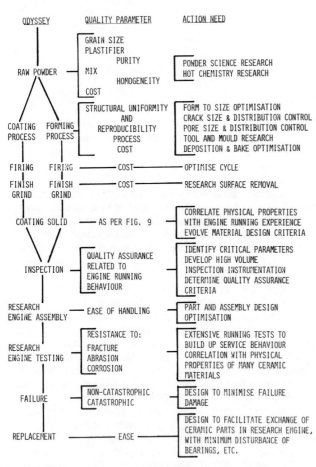

FIG. 10

840434

Advanced Adiabatic Diesel Engine for Passenger Cars

R. R. Sekar and R. Kamo
Cummins Engine Co.
Columbus, IN
J. C. Wood
NASA Lewis Research Center
Cleveland, OH

ABSTRACT

This paper presents the results of an analytical study to identify the essential features of a futuristic engine for cars. A combination of several advanced features in one engine package results in a dramatic increase in the fuel economy of the car while maintaining all other essential features at comparable levels to current cars on the market. This advanced adiabatic diesel (AAD) engine is expected to give 78.8 mpg in the Federal Combined Driving cycle in a 3,000-pound vehicle. This compares to 37.7 mpg and 30.0 mpg obtained in current diesel and gasoline engine powered cars, respectively. The study identified the research and development efforts needed to bring this concept to fruition and concluded that an aggressive 10-year program will result in production availability of these cars.

SINCE THE EARLY 70's, automotive manufacturers and other researchers have actively been seeking ways to reduce fuel consumption and exhaust emissions. Government efforts have been concentrated on the gas turbine and Stirling engines, while industry has been concentrating on the reciprocating I.C. engine. Industry efforts have been primarily directed toward the stratified charge engine with an increasing amount of effort on the diesel.

Research in the adiabatic diesel engine concept has been in progress for several years at Cummins (1,2,3)* with the sponsorship of Tank-Automotive Command (TACOM) of the U.S. Army. This research, however, is concentrated on heavy duty applications where the diesel engine already enjoys a dominant place. Yet several concepts and innovations that have resulted from the TACOM/Cummins program are thought to be equally useful in light-duty automotive applicatins. Hence, the Department of Energy, through NASA Lewis Research Center, sponsored a study at Cummins to study the feasibility of applying these adiabatic concepts to the light-duty diesel.

The objectives of the study were: (1) to identify the features of the advanced adiabatic diesel (AAD) engine that will give the best fuel economy in a 3,000-pound vehicle, (2) to identify the long lead time R&D areas that should be addressed in order to bring the selected concept to fruition, and (3) to propose the outline of a develoment program through the proof-of-concept phase.

The study consisted of several tasks which were used to identify the overall engine concept that would best meet the program objectives. The approach was to first look at various engine technologies and combine these into several candidate engines for further analysis. From these, one was then selected for detailed study.

Research in the adiabatic engine concept has been in progress for several years at Cummins (1,2,3)* with the sponsorship of Tank-Automotive Command (TACOM) of the U.S. Army. This research is concentrated on heavy duty applications where the diesel engine already enjoys a dominant place. Several concepts and innovations that came out of the TACOM/Cummins programs are thought to be equally useful in light duty diesel applications. Hence, this feasibility study was initiated at Cummins with the sponsorship of the Department of Energy. The program was monitored by NASA Lewis Research Center.

---

*Numbers in parentheses designate references at the end of the paper.

The objectives of the study were (i) to identify the features of the advanced adiabatic diesel (AAD) engine that will give the best fuel economy in a 3,000-pound vehicle, (ii) to identify the long lead time R & D areas that should be addressed in order to bring the selected concept to fruition, and (iii) to propose the outline of a development program through the proof-of-concept phase.

FEATURES OF THE SELECTED ENGINE

The engine is of adiabatic design featuring insulated pistons, cylinder liners, cylinder head, intake and exhaust ports, exhaust manifolds, and valves. No water or other liquid cooling system is needed, resulting in complete elimination of the conventional cooling system components such as radiator, water pump, etc. In order to insulate the hot gas stream components, extensive use of ceramic coatings and monolithic components is necessary. To reduce frictional and parasitic losses, the engine is designed to operate without liquid lubricant also. Antifriction ceramic rollers are used for wrist pin, crankpin, and main bearings. Solid lubricants are provided for other surfaces, such as cams and valve guides, which are normally lubricated with oil. To improve multifuel capability and to achieve "fast" heat release, a spark assisted, direct injection, high swirl diesel combustion system is used.

The charge air system includes a high efficiency helical screw compressor and a ceramic expander. These two units are connected by belts to the crankshaft. Thus, a simple positive displacement compounding system is achieved. These helical screw machines have higher efficiencies compared to conventional turbomachines and have excellent part load efficiencies.

Table 1 gives the major specifications of the AAD engine.

Table 1 - Major Specifications of the AAD Engine

Rating	- 70 BHP at 3000 rpm
Engine Cycle	- 4 Stroke, DI, Diesel
Bore x Stroke	- 77mm x 77mm (3.03 in. x 3.03 in.)
Displacement	- 1.4 liters (87.4 cu. in.)
Number of Cylinders	- 4 in-line
Compression Ratio	- 14.0
Maximum Cylinder Pressure	- 200 Bars (3000 psi)
Length, Width, Height	- 621 mm x 589 mm x 479 mm (24.4 in. x 23.2 in. x 18.9 in.)
Approximate Weight	- 136 kg (300 lb.)
Maximum BMEP	- 16 Bars (240 psi)
Firing Order	- 1-3-4-2

PROJECTED FUEL ECONOMY

The engine performance analysis was conducted using the Cummins Diesel Cycle Simulator (DCS) computer program (4). This is a thermodynamic model of the diesel engine cycle that has been validated by experimental data on several heavy duty engines. Some modifications were made to analyze the various advanced technologies included in the AAD engine. Figure 1 gives the AAD engine performance curves. The predicted fuel map of the engine is presented in Figure 2. These performance levels are considered to be practically achievable. Analysis was done at several levels of technology implementation from state of the art to ideal combination of the advanced concepts (5). The performance levels shown in Figures 1 and 2 are less than "ideal" but significantly better than state of the art diesel engines.

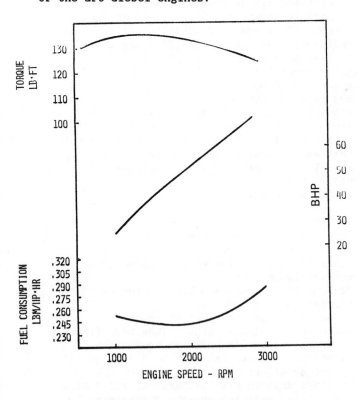

Fig. 1 - AAD Engine Performance Curves

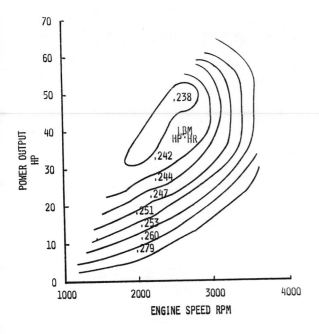

Fig. 2 - AAD Engine Fuel Maps

The impact of the AAD engine performance on vehicle fuel economy and acceleration characteristics was analyzed by Ford Motor Company research staff. The 1984 model Ford Tempo was chosen as the reference 3,000-pound vehicle. The computer program to analyze the vehicle performance in a Federal Driving Cycle was utilized extensively. The fuel map shown in Figure 2 was utilized in the vehicle analysis program, along with other vehicle related characteristics such as frontal area, drag coefficient, etc. In addition, a continuously variable transmission (CVT) was assumed to be part of the vehicle. The final results along with the expected performance using a state of the art IDI diesel are presented in Table 2.

Table 2 - Vehicle Performance Comparisons
(3,000 pound passenger car)

Power Train	Fuel Economy, MPG			0-60 MPH Accel. Time, Sec.
	City	Highway	Combined	
1. Baseline IDI Diesel	35.0	41.7	37.7	15.2
2. AAD Engine System	72.2	96.3	78.8	13.9

Additional fuel economy of about 3% is possible from vehicle aerodynamic drag coefficient reduction. A potential for thisdrag reduction exists due to the elimination of the radiator and the consequent lowering of the hood profile.

ADVANCED CONCEPTS INCORPORATED IN THE ENGINE

The significant performance gains projected are possible by combining several advanced concepts that are relevant to light duty applications. These concepts are discussed here briefly.

ADIABATIC CONCEPT - This concept relies on reducing the heat loss to the coolant by insulating the combustion chamber and other components exposed to the hot gas stream (1,2,3). This insulation is accomplished by high temperature, low conductivity ceramics. These could be in the form of monolithics, coatings, or composites depending on the particular component being insulated. In this study, an analysis was conducted to determine the effect of insulation on engine fuel consumption. When the engine is insulated, the brake specific fuel consumption (BSFC) improves. But the engine volumetric efficiency decreases. The net effect, shown in Figure 3, indicates that about 60-70% of total possible insulation is optimum.

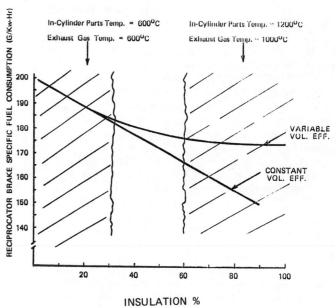

Fig. 3 - Effect of Insulation on Engine Fuel Consumption

While the adiabatic concept itself results in only a modest but measurable improvement in in-cylinder efficiency, the increase in available exhaust energy is large. Therefore, if a high efficiency recovery device is available downstream of the engine, the overall engine BSFC gains could be significant. In addition to the fuel economy improvement, the adiabatic concept results in complete elimination of the cooling system. This results in reduction of a substantial number of parts and cost.

SPARK ASSISTED DIESEL COMBUSTION SYSTEM - The combustion system chosen for this engine is a high swirl, direct injection, spark assisted diesel system. The spark assisted diesel has been tested and found to give excellent multifuel capability and improved engine performance and emissions characteristics (6). This combustion system is expected to provide a short duration heat release compared to conventional diesels. Analysis indicates that up to 18% improvement in BSFC is possible at part loads. The adiabatic and spark assistance concepts are expected to provide excellent starting characteristics for the engine.

HIGH CYLINDER PRESSURE - Peak cylinder pressures in state of the art heavy duty diesel engines are around 140 bars (2100 psi). The AAD design limit was chosen to be 200 bars (3000 psi). This limit is arbitrary and simply an estimate of evolutionary levels in the 1990's. Due to the novel features of the charge air system, the peak cylinder pressures remain high compared to conventional engines even at part loads. This feature is predicted to yield about 7% BSFC improvement.

COMPOUND SYSTEM - Turbocompounding of heavy duty diesel engines have been demonstrated (7). The turbocompound system discussed in (7) contains a conventional turbocharger, a power turbine, and a reduction gear train including a fluid coupling. For a passenger car application, it was felt that a similar system would be too complex and costly compared to the gains. Moreover, the car application is heavily weighed in favor of light load operation where turbomachinery efficiencies are poor. For these reasons, a positive displacement (PD) helical screw type compressor and expander were chosen for the AAD engine. Helical screw compressors have been widely used in industry for many years (8), and the expanders have been tested in several applications (9). Therefore, this novel feature of the AAD is a new application of known devices. The compounding is accomplished by connecting the compressor and expander by belts to the crankshaft. Table 3 illustrates the power gain of the PD compounding. At part loads, up to 34% of the needed power is provided by compounding. Also, the helical screw compressor and expander are predicted to have higher efficiencies compared to comparable size turbochargers. Figure 4 illustrates the efficiency curves used in this study.

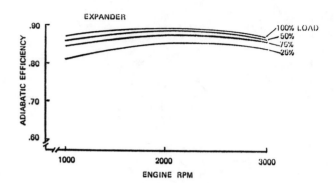

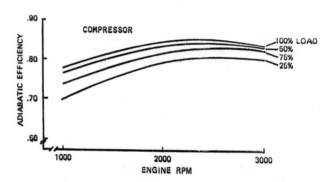

Fig. 4 - Efficiencies of Compressor and Expander

Table 3 - Work Split Between the Reciprocator and the Compound System

Load %	RPM	Reciprocator Power (kw)	Expander Power (kw)	% of the Recipr. Power
100	2500	52	13	25
	2000	42	9.6	23
	1000	20.5	4	19
75	2500	38	9	24
	2000	32	7	22
	1000	16	3	19
50	2500	26	8	31
	2000	21	6.6	31
	1000	10	2.8	28
25	2500	19	6.4	34
	2000	14.5	4.9	34
	1000	6.7	1.9	28

LOW FRICTION CONCEPTS - Oilless running of an engine is currently being investigated for heavy duty advanced adiabatic engines in the TACOM/Cummins program. The conceptual engine is shown in Figure 5. Extensive use of ceramic roller and needle bearings and solid lubricants are envisioned to implement this concept. In addition to reducing engine friction, elimination of oil also solves a difficult problem in the adiabatic engine — oil breakdown at high component temperatures. It is estimated that this concept would help reduce the mechanical friction loss by 60%, compared to a conventional diesel engine.

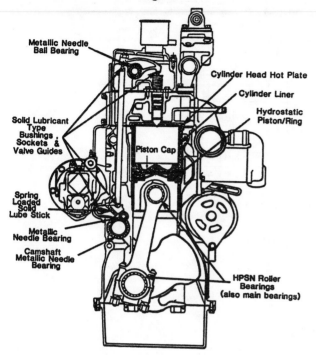

Fig. 5 - Low Friction Engine Concept

PREHEAT CONCEPT - Preheating of charge air with available exhaust thermal energy (regeneration) is a well known feature in gas turbines. This method of efficiency improvement could not be applied to conventional diesel engines due to metal component temperature limits. With the extensive use of ceramics, however, in the AAD, the regeneration principle becomes practical in diesel engines. This concept is described in detail in (5). In combination with the high efficiency compound system, preheating provides remarkable improvement in engine efficiency at part loads. The usefulness of this concept decreases when the duty cycle of the engine is more heavily weighed toward rated power where aftercooling becomes more effective.

PROJECTED EMISSIONS

There are no reliable analytical models available to predict the emissions characteristics of adiabatic engines. However, based on data available from heavy duty engine programs at Cummins (10), the key emissions characteristics were projected for the AAD. For this study, targets for HC, CO, NOx, and particulates were 0.4, 3.4, 1.0, and 0.2 g/mile. Projections, based on available emissions data on heavy duty adiabatic engines and predicted in-cylinder conditions for this engine, yield HC, CO, NOx, and particulate emissions of 0.13, 1.30, 1.65, and 0.18 g/mile. Even though the NOx

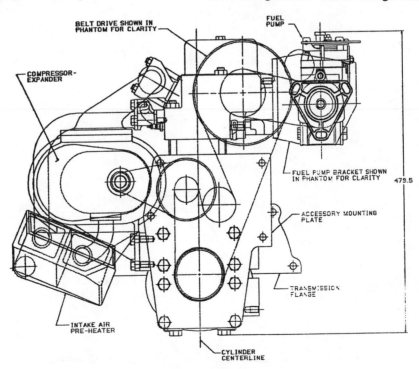

Fig. 6 - AAD Engine Layout - Front View

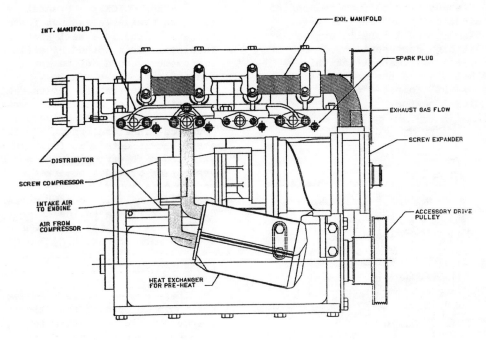

Fig. 7 – AAD Engine Layout – Side View

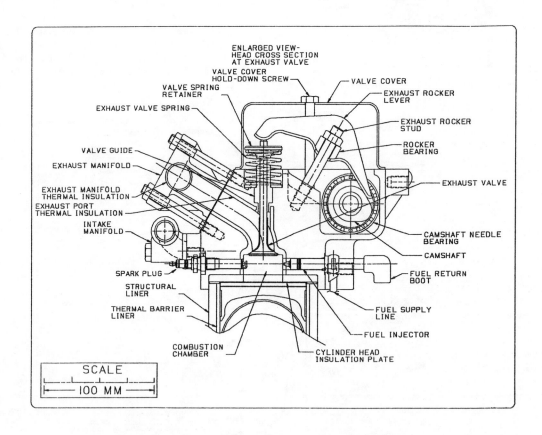

Fig. 8 – AAD Engine Cross-Section of Combustion Chamber

projection does not meet the target, it is comparable to current diesel engines. It should be emphasized that more experimental data on adiabatic diesel engines will have to be obtained before any definite conclusions could be drawn. It is true that higher in-cylinder temperatures of the AAD tend to increase NOx emissions. But, the expected reduction in ignition delay will reduce NOx. The net effect is at present unknown.

STRUCTURAL DESIGN FEATURES

Design and layout drawings of the AAD engine were made using the Computer Aided Design (CAD) system. Figures 6 and 7 are two overall views of the engine. The front view, Figure 6, shows the compressor and expander sized for this engine connected by a belt to the engine crankshaft. The air preheater shown is sized and located for easy installation in the engine compartment. The fuel pump is driven off of the overhead cam. Figure 7 shows a side view of the spark plugs and the distributor. The cross section through the combustion chamber, shown in Figure 8, illustrates spark plug and the fuel injector positioning. The valve actuation mechanism is conventional.

The piston design in an oilless, adiabatic engine is complicated and poses a difficult problem primarily because of side thrust. A novel piston design with a rolling crosshead is incorporated in the engine design as one possible solution. However, experimental verification and careful piston design are needed during hardware development.

HEAT TRANSFER ANALYSIS

The major structural problems, if any, are likely to arise from the high component temperatures caused by the adiabatic concept. Three dimensional finite element models (FEM) were built for the head, piston, and liner to assess the feasibility from the standpoints of component temperatures and thermal stresses. Figure 9 illustrates the cylinder head FEM as an example. Figures 10, 11, and 12 illustrate the typical temperature plots of the head, piston, and liner. These temperature levels are significantly higher than conventional engine components but well within the safe limits of the ceramics and other materials chosen for this engine. The stress levels, both thermal and mechanical, are acceptable. The critical stress-to-material strength ratios of the major components are close to heavy duty diesel values for 250,000 miles in service durability. However, statistically, significant number of ceramic parts have not been tested for durability. In the AAD engine, typical car durability of 100,000 miles is deemed to be achievable.

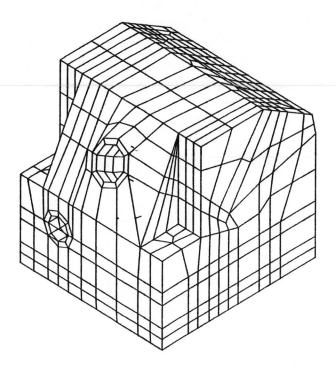

Fig. 9 - FEM of AAD Head

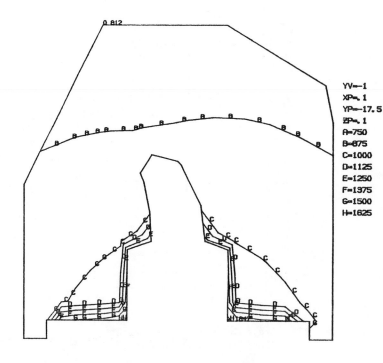

YV=-1
XP=.1
YP=-17.5
ZP=.1
A=750
B=875
C=1000
D=1125
E=1250
F=1375
G=1500
H=1625

Fig. 10 - Temperature Plots of Head at Steady State Rated Power

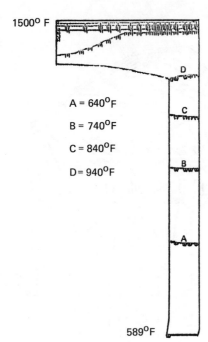

1500° F

A = 640°F

B = 740°F

C = 840°F

D = 940°F

589°F

Fig. 11 - Example of Piston Isotherms
at Steady State Rated Power

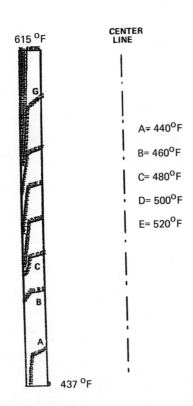

615 °F

CENTER
LINE

A = 440°F

B = 460°F

C = 480°F

D = 500°F

E = 520°F

437 °F

Fig. 12 - Typical Liner Isotherms at
Steady State Rated Power

COST

Cost estimates of this AAD engine have to be speculative due to the length of time involved to reach production and to the uncertainties involved with the many novel features. A rough estimate of the cost indicates the AAD to be twice as much as the conventional diesel. However, on a 100,000 mile ownership life cycle cost basis, the AAD powered car will save a significant amount of money compared to gasoline and conventional diesel powered cars. Several design related cost reduction potentials have already been identified to make the initial cost equal to current diesels and improve the life-cycle-cost savings further.

CRITICAL RESEARCH AREAS

The pacing item has been already recognized to be ceramic components - design, mass production, cost reduction, etc. This area is already being emphasized by several organizations. Additionally, the specific areas that need experimental research in order to make the AAD powered car a reality are:

- Positive Displacement Compounding System
- Preheat Concept
- Piston Design for Oilless, High Temperature Operation
- Combustion System with Spark Assist
- Oilless Bearing

These are, in fact, generic technologies that would benefit light, medium, and heavy duty applications of an advanced diesel engine. Development of these technologies should proceed simultaneously with the development of the AAD powered car. An aggressive schedule and resource commitment in several parallel efforts could lead to production in 10 years from the date of commencement of the program.

SUMMARY

This study indicates that the advanced adiabatic diesel engine powered car would give a dramatic increase in fuel economy while maintaining all the other essential features at comparable levels to current cars. The engine concept offers the potential to the consumer to save money over the life cycle and offers the country one of the best means of reducing petroleum imports. The engine includes several novel, "high tech" features which make it a high payoff, high risk proposition. It is a very practical engine which could be brought to production with firm commitment and an aggressive schedule.

ACKNOWLEDGEMENT

The authors would like to acknowledge the contributions of several members of the Cummins project team and Ford Motor Company's research staff. In addition, Cummins and NASA Lewis Research Center management support and the financial support from the Department of Energy are sincerely appreciated.

REFERENCES

1. R. Kamo and W. Bryzik, "Adiabatic Turbocompound Engine Performance Prediction", SAE Paper 780068, Feb., 1978.

2. R. Kamo and W. Bryzik, "Cummins/TARADCOM Adiabatic Turbocompound Engine Program", SAE Paper 810070, Feb., 1981.

3. W. Bryzik, "TACOM/Cummins Adiabatic Engine Program", Department of Energy Automotive Technology Development Contractors Coordination Meeting, October, 1982, SAE Publication P-120.

4. A. S. Ghuman, M. A. Iwamuro, and H. G. Weber, "Turbocharged Diesel Engine Simulation to Predict Steady State and Transient Performance", ASME Publication No. 77-DGP-5.

5. L. Tozzi, R. Sekar, R. Kamo, and J. C. Wood, "New Perspectives for Advanced Automotive Diesel", 18th Intersociety Energy Conversion Conference, Orlando, Florida, August, 1983.

6. R. Kamo, "Synfuel Modified Diesel", Final Report to U.S. Department of Energy and Battelle Northwest Labs, Contract No. B-A0763-A-P, May, 1982.

7. J. L. Hoehne, and J. R. Werner, "The Cummins Advanced Turbocompound Diesel Engine Evaluation", SAE Publication P-120.

8. "Production Design of a Modern, Axial Flow Positive Displacement Rotary Compressor", Paper No. 59-DGP-3, ASME, April, 1959.

9. R. F. Steidel, Jr., H. Weiss, and J. F. Flower, "Performance Characteristics of the Lysholm Engine as Tested for Geothermal Power Applications in the Imperial Valley", ASME, Journal of Engineering for Power, Vol. 104, p. 231, January, 1982.

10. R. Kamo, L. Tozzi, and R. Sekar, "The Light Duty Diesel of the Future", Automotive Engineering, Oct., 1983.

Exploratory Study of the Low-Heat-Rejection Diesel for Passenger-Car Application

Donald C. Siegla and Charles A. Amann
Engine Research Department
General Motors Research Laboratories

ABSTRACT

Eliminating the conventional liquid cooling system of a diesel engine to conserve energy normally rejected to that heat sink offers promise as a means for improving fuel economy. Such low-heat-rejection (LHR) diesels have generally been advanced for heavy-duty vehicles. In this study, application of the concept is analyzed for a light-duty indirect-injection diesel of the type used in passenger cars. The naturally aspirated LHR diesel is found to offer no fuel economy advantage, principally because of the deteriorated volumetric efficiency arising from hot cylinder walls. It is found that most of the energy conserved by deleting the cooling system is diverted to the exhaust gas. Methods examined for recovering the lost volumetric efficiency and/or harnessing the increased energy content of the exhaust include supercharging, adding a bottoming cycle, and combining the diesel with turbomachinery. The latter option is judged superior for the passenger-car application.

THE CONCEPT OF HEAT CONSERVATION looms as a bright prospect on the horizon for future diesel engines. In its ultimate form, this concept eliminates the traditional water/glycol cooling system, along with the space, cost and maintenance requirements which attend that system. Such a power plant frequently has been called an "adiabatic" engine. However, that is in fact a misnomer because interchange of heat between the cylinder gases and their confining walls, although decreased substantially, is not eliminated as the term "adiabatic" implies. In the present study an engine incorporating heat conservation is instead called a "low-heat-rejection" (LHR) engine.

The LHR approach is ill-suited to the conventional gasoline engine because high cylinder-gas temperature promotes combustion knock, but it is ideal for the diesel engine. Among the positive attributes of the LHR diesel, beyond elimination of a cooling system that includes a coolant pump, radiator, and cooling fan, are the potential for improved fuel economy as a result of decreased heat rejection, and a reduction in the ignition-delay period. Shorter ignition delay may allow the use of a lower compression ratio, with a consequent decrease in motoring power that could contribute to a higher mechanical efficiency, hence better fuel economy.

It would be unfair to overlook some new concerns associated with the LHR diesel. First, the higher operating temperature imposes more stringent demands on the lubricant. Specially formulated synthetic oils have been used in short-term research operation. Looking ahead, such fresh approaches as the use of dry lubricants and/or air lubrication [1]* are under consideration. Second, in applications where driver/passenger comfort is important, as in the passenger car, innovation is needed to accomplish effectively the heater/defroster function previously handled with the engine cooling system.

The outlook for exhaust emissions is uncertain in the LHR diesel. Diesel automobiles presently marketed in the U.S. are operating under a waiver of the 1.0 g/mi NOx standard that allows levels up to 1.5 g/mi in most vehicles. By statute, those waivers are set to expire at the end of the 1984 model year. The passenger-car diesel also faces future particulate regulations that are generally agreed by manufacturers to lie beyond the capability of current engine technology. The use of fuel additives and specially blended fuels [2], and of exhaust particulate traps [3] have all been investigated, but have so

---

*Numbers in brackets designate references found at the end of this paper.

far fallen short of a practical production solution. The effect of LHR on emissions must be watched with care because an engine that offers better fuel economy at the expense of noncompliance with emission standards is clearly unacceptable.

The technical development that has recently kindled interest in the LHR diesel is the progress being shown with structural ceramics and ceramic coatings. Although this progress has been sufficient to facilitate demonstration of LHR diesels [4], such issues as durable component design, large-scale manufacture, production quality, inspection techniques, and cost remain to be resolved. Meanwhile studies aimed at assessing the potential payoff in engine performance and/or fuel economy upon achieving success in these areas is in order.

Published analyses have been quite limited in number. Kamo and Bryzik reported results from the full-load evaluation of a direct-injection LHR diesel for military use [5]. Wallace et al. examined several alternative LHR configurations for a direct-injection truck engine and confirmed the desirability of turbocompounding [6]. Way and Wallace then analyzed aspects of turbocompounding such an engine [7], but only at full load. This is appropriate for the truck application, where engine usage is characterized by considerable operation at high loads and near rated speed. The analysis by Tovell was also for a direct-injection truck engine [8].

In contrast to the heavy-duty truck application, in the passenger car the engine is operated primarily at part load, and at speeds well below that for maximum power. Furthermore, although the superior fuel economy of the direct-injection diesel is also sought in the passenger car, to date that type of diesel has been unable to satisfy U.S. passenger-car emission standards. Consequently, all passenger-car diesels presently marketed are of the indirect-injection type, with the fuel being injected into an antechamber that is joined by a passageway to the main chamber above the piston.

What distinguishes the present study from earlier analyses cited above is that this one is aimed at the passenger-car application and therefore considers operation at part load, and that it explores the LHR approach in an indirect-injection configuration. It uses two types of models -- the phenomenological model and the system model. The former is applied only to the engine cylinder and deals with the various phenomena comprising the cycle on an individual basis, combining the resulting submodels to predict the performance parameters that characterize the naturally aspirated engine *in toto*. As the study proceeds, it proves desirable to consider the addition of such features as a supercharger or a bottoming cycle. In the analysis of the resulting engine system, the constituent components are treated as separate entities, without concern

about the detailed phenomena occurring within them. This is the system model.

The phenomenological model, which is discussed later, is not a consummate model. The mathematical treatment given individual phenomena is necessarily restricted by computer capability, and the submodels also reflect the current imperfect understanding of in-cylinder events. Therefore, extreme precision in any performance projections made in this study is an unreasonable expectation. The value of this exercise is rather in pointing to what general characteristics should be anticipated as engine heat conservation is pursued.

The paper begins with some considerations fundamental to the concept of heat conservation. This is followed by a discussion of the phenomenological engine-cylinder model and its application. Next, results are presented for the naturally aspirated LHR engine. Both the use of an engine-driven supercharger and the addition of a bottoming cycle are then considered in turn. Finally some observations are offered regarding the combination of the LHR diesel with turbomachinery.

Symbols used in the equations scattered throughout the text are defined as they are introduced. For the convenience of the reader, these definitions are summarized in a nomenclature list that appears just ahead of the appendices.

THE FUEL-ECONOMY PROSPECT

The first and second laws of thermodynamics provide a foundation on which to base a fundamental understanding of the prospect for fuel economy, as well as certain limitations inherently imposed.

FIRST LAW - Application of the first law of thermodynamics dictates that all of the energy put into the engine through combustion of fuel must ultimately appear in one of three places -- (1) as work on the crankshaft, (2) as heat rejected either to the coolant, the lubricant, or directly to the environment, or (3) as energy exiting in the exhaust stream. The objective of maximizing energy to the first category -- work output -- then must come at the expense of a reduction in energy appearing in one or both of the other two categories. Of course the aim of the LHR engine is to decrease the fraction of the fuel energy going to the second category. Nearly all of the energy in the second category normally appears in the coolant, but if the coolant is eliminated as a heat sink, that energy may be redirected not only to shaft work, but alternatively to the lubricant, to radiation and convection from the engine skin, and/or to the exhaust stream in the form of increased gas temperature. (The exhaust stream may also contain chemical energy in the form of products of incomplete combustion, but in an engine meeting passenger-car emission standards, that involves a trivial fraction of the original fuel energy.)

In Fig. 1 the full-load power of a typical passenger-car diesel is plotted against engine speed. Within the engine operating range, contours of constant thermal efficiency, i.e., the fraction of fuel energy converted to shaft work, are also shown. The road-load requirement of a passenger car is superimposed. The load is clearly too low to utilize the efficiency capability of the engine to best advantage. The road-load level is set that low on the engine map in order to provide adequate power reserve for uphill operation, passing maneuvers, freeway merging, etc.

In Fig. 2 the same full-load curve is repeated, along with the road-load curve, but now the contours show the fraction of the fuel energy in the second category, which is principally heat to the coolant. Whereas the energy fraction to the output shaft in Fig. 1 was maximum near full load, the energy fraction to the coolant is seen in Fig. 2 to reach a maximum at no load. The road-load requirement lies in a region of high fractional heat rejection to the coolant. Its recovery on the output shaft would attractively complement the low road-load thermal efficiency, relative to the maximum available from the engine, of the light-duty passenger car.

SECOND LAW - While the first law defines the options for redistribution of energy in a LHR engine, the second law restricts the nature of that redistribution according to when in the engine cycle the conservation of rejected heat occurs. Applying to an indirect-injection engine the computer simulation that is described later in this paper, the heat rejected to the coolant during each of the four strokes comprising the cycle is listed in Table 1 as a fraction of fuel-energy input at 1000 r/min for both a high and a low engine load.

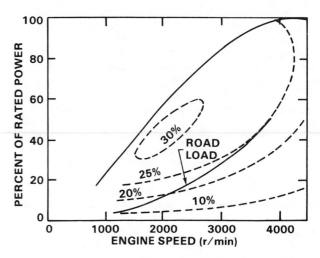

Fig. 1 - Typical engine map showing contours of constant brake thermal efficiency, with road-load operating line superimposed

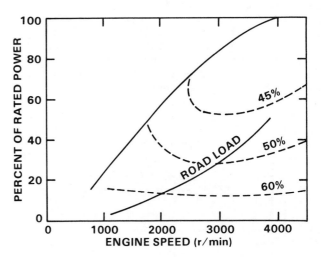

Fig. 2 - Typical engine map showing contours of heat rejection as a percentage of fuel input energy, with road-load operating line superimposed

TABLE 1

$$\left(\frac{\text{Heat rejected to coolant}}{\text{Fuel-energy input}}\right)$$

Engine speed	1000 r/min	1000 r/min
Engine load	Heavy	Light
Fuel-air ratio	0.0518	0.0155
(Coolant heat/fuel energy)		
Intake stroke	-0.012	-0.024
Compression stroke	0.041	0.150
Expansion stroke	0.218	0.297
Exhaust stroke	0.010	0.008
	0.257	0.431

It is seen from Table 1 that in the conventional engine the cylinder gas is heated slightly by the cylinder walls during the intake stroke and in turn rejects a comparatively small fraction of the fuel energy as heat to the cylinder walls during the exhaust stroke. The energy exchanged during these strokes is small compared to what is transferred during the compression and expansion strokes, however.

During the early part of the compression stroke the cylinder gas continues to absorb heat from the cylinder walls, but the direction of heat exchange reverses as the cylinder-gas temperature is increased by compression, and by top dead center (TDC) of the compression stroke the net transfer of heat has been out from the gas. Although it might appear at first glance from Table 1 that the net heat rejected during compression is substantially greater at light load than at heavy load, the heat rejected to the coolant during compression is actually nearly independent of load. The disparity in _fractional_ heat rejection listed in Table 1 occurs because of the considerably lower fuel per cycle at light load, which appears as the denominator of the fraction.

In a LHR diesel without a liquid cooling system the outflow of heat from the cylinder gas would be sharply curtailed by the hotter-running cylinder walls. This might at first seem a thermodynamic disadvantage to the engine cycle because compression work is decreased by the concurrent rejection of heat. In fact, if heat rejection in the standard engine could be increased enough to make the process isothermal, then at a compression ratio typical of the passenger-car diesel, compression work would be approximately halved.

The fallacy of focusing narrowly on the compression stroke in that manner is illustrated in Fig. 3, where pressure-volume diagrams are overlaid for an air-standard diesel cycle, which features adiabatic compression, and a similar cycle with the same heat addition, but with an isothermal compression process replacing adiabatic compression. The crosshatched area between the compression curves indicates the compression work saved by increasing, rather than conserving, heat transferred during compression. However, this gain is more than nullified by the oppositely crosshatched work area lost during combustion and expansion as a result of the lower compression pressure accompanying isothermal compression. Thus the heat conservation of the LHR diesel pays off thermodynamically during the compression process unless it causes counteracting adverse effects in the rest of the cycle.

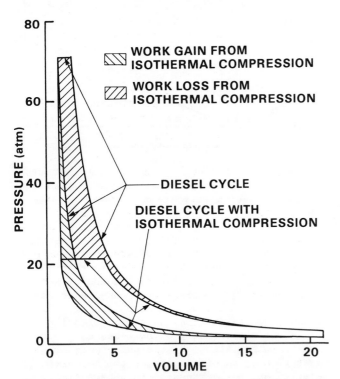

Fig. 3 - Ideal diesel cycle compared to same cycle with isentropic compression replaced by isothermal compression, constant heat added

Not surprisingly, most of the heat rejected to the coolant is seen in Table 1 to be lost during the expansion stroke, which normally encompasses combustion in the indirect-injection diesel. Unfortunately, even if that heat transfer were to be totally eliminated in a LHR engine, it would be impossible to recover all of that energy as indicated work in the cylinder gas. This is obvious once it is recognized, for example, that if the increment of heat rejected at bottom dead center (BDC) were conserved, none of it could be converted to work because the piston has already completed its power stroke.

The middle curve in Fig. 4 portrays the instantaneous rate of energy rejection to the coolant during the expansion stroke in an indirect-injection diesel running at 1000 r/min and a fuel-air ratio of 0.0518. The difference between the cylinder-gas temperature and the confining chamber walls, which drives the transfer of heat, is high near TDC and falls as the cylinder volume expands. In contrast, the exposed wall area through which the heat is transferred grows from a minimum at TDC to a maximum at BDC. The net result of these opposing variations, acting together with a changing average heat transfer coefficient, is seen from Fig. 4 to cause a rise in heat transfer rate from a substantial level at the end of compression to a maximum, in this case at about 20 deg ATDC, followed by a drop to a very low level at BDC.

To determine how much of this heat lost to the coolant could be converted to indicated work if the heat could be completely conserved, it is necessary to multiply each point on the heat-rejection curve of Fig. 4 by its corresponding work factor. Work factor, developed in Appendix A, is a function of the cylinder expansion ratio remaining at the time an increment of heat is lost. Plotted at the top of Fig. 4, it is seen to fall from a maximum of 0.63 for this engine at TDC to zero at BDC.

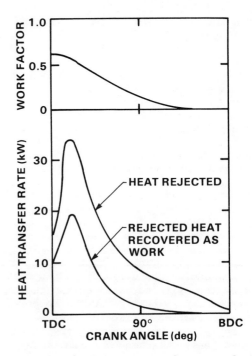

Fig. 4 - Typical heat transfer rate during expansion stroke and effect of second law on portion recoverable as work

The lowermost curve of Fig. 4 is the result of multiplying the instantaneous heat rejection by the work factor. The area under this curve represents the energy that could theoretically be converted to indicated work if the engine could be made truly adiabatic during the expansion stroke. It represents only about one-third of the area under the curve of instantaneous heat rejection during expansion. From the first law, with complete heat conservation the remaining two-thirds would exit with the exhaust stream as increased exhaust-gas temperature.

DESCRIPTION OF ENGINE MODEL

The phenomenological model of the diesel-engine cylinder used in this study evolved from the cycle-simulation code developed at MIT under a grant from General Motors [9]. Additional refinement can be anticipated as experimental research provides the understanding on which to base further improvements. Validation studies of the original code, conducted using a single-cylinder indirect-injection engine, have been reported [10]. Additional evaluation of the simulation was conducted by comparing nitric oxide histories during the expansion stroke, as calculated by the simulation, with measurements made in a single-cylinder engine at the University of Minnesota under another General Motors grant [11].

The cycle simulation uses a stochastic representation of the fuel-air mixing process in an indirect-injection diesel engine. This approach has appeal because high-speed movies taken in the prechamber have revealed a degree of randomness to the combustion pattern from one cycle to the next.

The prechamber and the main chamber are modeled as partially stirred reactors coupled by a connecting passageway. At intake-valve closing the contents of the two chambers are comprised of over 1000 equal-mass increments of air and residual products.

Up to the point of intake-valve closing, inflow into the main chamber is modeled as a quasi-steady flow process across a variable-area orifice, the intake valve. Although the intake manifold is assumed adiabatic, heat transfer takes place between the charge and the main-chamber walls once the air charge is ingested into the cylinder. Heat interchange between the gas and the walls is computed using a Nusselt-Reynolds number relationship that was developed for turbulent pipe flow. The velocity used in the Reynolds number is a characteristic mean velocity derived from conservation of kinetic energy.

At intake valve closing, the trapped mass in the cycle is established. Thereafter the piston forces mass from the main chamber into the prechamber. The mass flowing into the prechamber is governed by the pressure drop across the effective area of the connecting passageway assuming quasi-steady one-dimensional flow. Beginning at the start of injection,

fuel is introduced into the prechamber in increments of mass equal to the mass of an air element until the specified amount of fuel has been injected. Fuel elements persist as liquid until they have been evaporated.

The general practice in published cycle simulations is to prescribe the rate of energy release during combustion *a priori*. In contrast, in this simulation the energy-release rate is calculated by the simulation as a function of certain input parameters. In each chamber, elements of fuel and air mix with each other on a random basis. At a given time, two random elements are mixed and then divided. If a resulting element contains both fuel and air in a proportion between prescribed rich and lean limits and has also fulfilled a computed ignition delay time, then it burns. Burned elements can mix randomly with unburned elements or with elements of fuel vapor and react according to chemical equilibrium considerations.

The rate of mixing of elements in each chamber is proportional to the instantaneous turbulence kinetic energy in that chamber. Turbulence kinetic energy is computed from the characteristic mean velocity in the chamber through a turbulence-dissipation expression. A typical time interval between mixing events occupies less than 0.01 deg of crank angle.

During combustion the formation of nitric oxide in each element is computed according to extended Zeldovich kinetics. In the original simulation [9], soot formation was modeled according to equilibrium chemistry primarily as a convenience in formulating the code. Although equilibrium chemistry has been considered in the absence of more realistic models in order to gain insight into the possible influences of various factors on soot formation in engines [12], the formation of carbonaceous soot during combustion is recognized to be a nonequilibrium process. From the results of an international symposium on the formation of particulate carbon during combustion [13], it was concluded that no more suitable technique for modeling soot formation in an engine was presently available than the experimentally based Arrhenius expression derived by Khan et al. [14]. The equilibrium submodel was therefore replaced with this one. Soot oxidation is computed using the expression of Nagle and Strickland-Constable [15]. Beyond the validation studies previously mentioned, additional engine experiments were run to check the validity of these emissions calculations. Fair agreement was found between measured and calculated emission indices for oxides of nitrogen (EINOx) and particulate carbon (EIC) in the engine exhaust. The agreement between measured and calculated indicated mean effective pressure (IMEP) was good.

The heat-transfer submodel of the original simulation required manipulation to represent the low-conductivity wall without liquid coolant that typifies the ultimate LHR diesel.

In the standard engine, heat leaves the cylinder gas through three thermal resistances in
series: the gas-side film in the cylinder,
the solid wall of the combustion chamber, and
the liquid film on the coolant side of the
wall. In simulating the LHR configuration,
the correlation for the gas-side film was
retained, but the other two resistances were
lumped into a single, constant conductance
ratio (U*). The definition of U* is detailed
in Appendix B. A value of 0.2 was chosen for
this study, corresponding to an 80 percent reduction in conductance of the chamber wall and
its outside film.

With U* and the underhood temperature
(82°C) assigned, the representative main-
chamber inner-wall temperature for the LHR
diesel, which is required input for the simulation, can be estimated from the experimental
results of Alkidas and Cole [16] using the
technique outlined in Appendix B. Corresponding
prechamber and passageway wall temperatures
were estimated by assuming their elevations
above the temperature of the main-chamber wall
would remain the same as measured for the
baseline condition. The calculated wall
temperatures for the LHR engine are compared
with the measured temperatures for the baseline
engine at speeds of 1000, 2000 and 3000 r/min
in Fig. 5. The peak cylinder-wall temperatures
of 800-825 K compare with a ring/liner interface temperature at the location of top-ring
reversal of 800-925 K anticipated in a fully
developed heavy-duty LHR diesel [18].

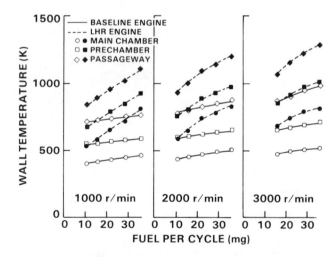

Fig. 5 - Wall temperatures versus fueling
rate at three speeds, LHR and baseline
engines

The simulation provides indicated work
and pumping work developed in the cylinder,
but it is brake work at the output shaft that
is of interest. This raises the perplexing
question of how to treat mechanical friction.
Even in a conventional engine its determination
during operation is a difficult task that is
seldom undertaken. Motoring torque is often

accepted as a reasonable substitute even
though during motoring the gas forces acting
on the piston are different from their values
when firing, the piston and cylinder-wall
temperatures are lower, and the oil viscosity
is increased. Piston friction, the major
contributor to the mechanical-friction load,
also depends on piston mass, piston-to-bore
clearance, ring tension, and the general level
of lubricant viscosity, all of which may be
changed in converting from a conventional to a
LHR configuration. Given this situation,
mechanical-friction mean effective pressure was
based on motoring test measurements from a 5.7-L
indirect-injection production diesel. Because
the compression ratio in the LHR engine was
reduced from the production value of 22.5 to
19.2 to take advantage of the higher gas temperatures, the values of friction mean effective pressure (FMEP) determined by motoring the
production engine were lowered appropriately
[19]. Values used were:

1000 r/min	-	150 kPa
2000 r/min	-	175 kPa
3000 r/min	-	200 kPa

How these FMEPs should be varied with load is
another question. Taylor and Taylor show a
modest increase in FMEP with increasing cylinder
pressure [20], but in one diesel engine Millington and Hartles found that friction actually
decreased slightly as load was increased [19].
In this study, friction at a given speed was
assumed invariant with load except where specified otherwise.

Parasitic engine loads deserving special
attention include the radiator fan and the
coolant pump, both of which would disappear in
a LHR engine along with the liquid cooling
system. With respect to the cooling fan, it is
helpful to recognize that fan power requirement
tends to vary with the cube of speed. With the
engine in a typical light-duty passenger car
normally running below half its rated speed
much of the time, this means that the engine-
driven fan is often absorbing less than a tenth
of its rated-speed power. In fact, in front-
wheel-drive cars with electrically driven fans,
the fan is not even operated during many
driving conditions. This stands in marked
contrast to the heavy-duty long-haul truck,
which normally operates nearer the engine rated
speed and consequently carries a higher average
cooling-fan power requirement. This suggests
that the beneficial effect of eliminating the
cooling-fan load in the heavy-duty engine, on
which most of the LHR development effort has
been focused to date, would be substantially
diminished in the passenger-car application.
(If the standard heavy-duty engine is equipped
with a thermostatically controlled fan, then
the benefit from eliminating the fan in the LHR
heavy-duty engine would be reduced as well.)
Elimination of the comparatively small coolant-
pump load is an undeniable plus. On the other
hand, the possible need for a small oil-cooler

fan is foreseen in the LHR engine, but that is of course dependent on the lubricant developed for the application. In the face of all these considerations, the parasitic loads of fans and the coolant pump were excluded from this study unless otherwise noted.

The specifications for the engine that formed the basis of this study are given in Table 2. As noted in the ensuing discussion, the displacement of this baseline engine was sometimes scaled up or down to make realistic comparisons of various cycle options. A cross section through the baseline engine is presented in Fig. 6.

TABLE 2 - Engine Specifications

Bore	103.0 mm
Stroke	85.6 mm
Connecting rod length	223.5 mm
Displacement*	0.718 L
Prechamber volume*	0.0147 L
Compression ratio	19.2
Intake valve opens	6 deg BTDC
Intake valve closes	38 deg ABDC
Exhaust valve opens	64 deg BBDC
Exhaust valve closes	17 deg ATDC

*per cylinder

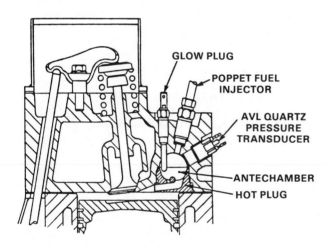

Fig. 6 - Cross section of baseline-engine cylinder

NATURALLY ASPIRATED ENGINE

In simulating the effect of an 80 percent reduction in conductance of the chamber wall and its outer film, the first question of interest is to what degree the percentage of fuel energy rejected from the cylinder gas is influenced. In Fig. 7 those percentages are plotted against overall fuel-air equivalence ratio, i.e., load, for both the baseline engine specified in Table 2 and an LHR version of that engine (U* = 0.2). Reviewing the curves for the baseline engine first, the potential for

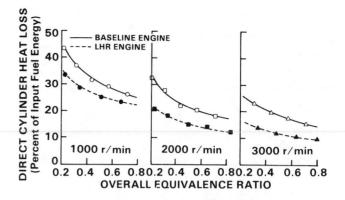

Fig. 7 - Direct cylinder heat loss versus load at three speeds, LHR and baseline engines

the LHR approach appears particularly attractive in light-duty use because the rejected heat is the greatest percentage of the fuel energy at low speeds and light loads. However, the LHR curves show that the fraction of that rejected heat that is actually conserved in the LHR engine is the least at 1000 r/min and the greatest at 3000 r/min.

The effect of this altered heat-rejection pattern on indicated thermal efficiency is seen in Fig. 8. The heat conserved at 1000 r/min has essentially no effect on indicated thermal efficiency at 1000 r/min, but it does cause a gain at 2000 r/min, and to some extent at 3000 r/min.

It is clear from the first law that any of the reduced coolant loss shown in Fig. 7 that does not appear as increased efficiency in Fig. 8 must escape from the engine with the exhaust gas. The increased energy in the exhaust is manifested as higher exhaust gas temperature, as illustrated in Fig. 9.

The parameters plotted in Fig. 7-9 are important to understanding the thermodynamics of the LHR diesel, but the engine user is interested instead in brake power as a function

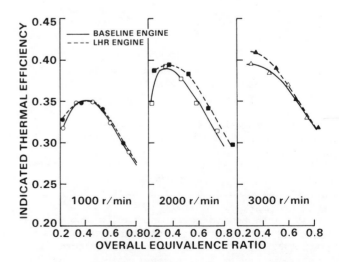

Fig. 8 - Indicated thermal efficiency versus load at three speeds, LHR and baseline engines

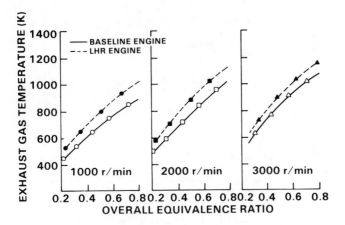

Fig. 9 - Cylinder exhaust gas temperature versus load at three speeds, LHR and baseline engines

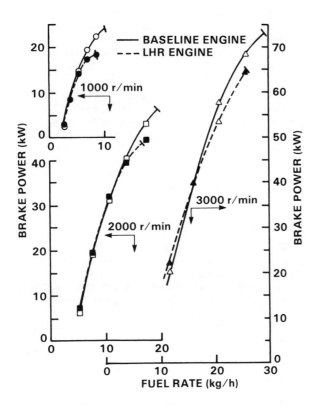

Fig. 10 - Brake power versus fueling rate at three speeds, LHR and baseline engines

of fuel rate, which is plotted for both the baseline and LHR engines in Fig. 10. (A different origin has been selected for the coordinate system at each speed to avoid excessive interference among the curves.) Each power curve is terminated at its upper extremity by a tick that marks an overall fuel-air ratio of 0.0555, selected to represent the full-load limit for both engines on the basis of experience with current indirect-injection diesels.

There are three important points to be noted from Fig. 10. First, the full-load power is lower in the LHR diesel -- by 12 percent at

3000 r/min, 14 percent at 2000 r/min, and 24 percent at 1000 r/min. Second, this increasing spread in full-load power with falling speed signifies a less favorable torque-curve shape for the LHR engine. Third, except for light load at high speed, where a passenger-car engine does not often operate, the naturally aspirated LHR engine shows no significant fuel economy advantage. This last point is apparent from comparing the power developed by each engine at a given speed and any arbitrarily selected fuel rate.

But this fuel-economy comparison between engines of unequal power capability is not realistic. A more reasonable comparison would result if the displacement of the LHR diesel were increased 12 percent so that it could match the output of the baseline engine at 3000 r/min. Of course this demands a 12 percent increase in fuel rate at any given speed and brake power in the LHR engine. The implications of such a scaling exercise are shown for 3000 r/min in Fig. 11. The solid curve for the baseline engine and the dotted curve for the LHR engine of equal displacement are repeated from Fig. 10. The fuel-economy advantage of the LHR engine below half load is demonstrated by its lower fuel rate at any specified brake power below 37 kW. The broken-line curve is the result of increasing the displacement of the LHR engine to match the full-load power of the baseline engine. (The fact that the full-load fuel rate of the scaled engine coincides with the full-load requirement of the baseline engine is conincidental.) Now the LHR engine

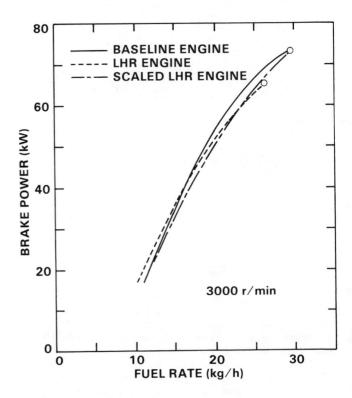

Fig. 11 - Brake power versus fueling rate at 3000 r/min, baseline engine, LHR engine, and LHR engine scaled up to match full-load power

has lost its fuel-economy advantage in the 25 to 50 percent load range, falling below the baseline engine by as much as 8 percent. Although not shown in Fig. 10, its penalty reaches a maximum of 3 percent at 2000 r/min and 5 percent at 1000 r/min.

CRITIQUE - The disappointing fuel economy calculated for the naturally aspirated LHR engine is not what one might at first expect and merits further examination. Three aspects of the calculation will be considered in greater detail: combustion timing, volumetric efficiency, and certain aspects of the heat-loss estimate.

Combustion Timing - In the calculations, the crank angle at start of injection at each speed was based on production practice and was not altered for the LHR engine. Given the hotter prechamber environment of the LHR engine, one might expect a shorter ignition delay, with timing being overadvanced as a result. Over the broad range of conditions investigated, however, the calculated change in ignition delay at a given speed never exceeded 2 crank-angle deg. This could be the result of an inadequate ignition-delay model, or it could be characteristic of an engine that has a short ignition delay in the first place, as was the case in the baseline engine.

It is also possible that a faster rate of combustion in the hotter cylinder environment would call for a later start-of-combustion timing. On the other hand, this need not be expected if the combustion is mixing-controlled.

The combined effect of ignition delay and combustion rate can be evaluated by calculating a timing curve, assuming the simulation itself is valid. A timing curve of indicated mean effective pressure (IMEP) versus injection advance relative to the baseline setting is presented in Fig. 12 for one speed and fuel rate. Although a slight improvement is indicated for a more advanced timing, the curves are essentially parallel. Hence the poor fuel-economy showing of the LHR engine is not the result of improperly selected combustion timing.

Volumetric Efficiency - On account of heat conservation in the LHR engine, some degradation in volumetric efficiency might be expected because heat transferred from the walls to the incoming air during intake decreases the density of the air trapped in the cylinder at intake valve closing. Consider the adverse effect of decreased volumetric efficiency (based on intake-manifold density) on brake thermal efficiency at fixed speed (N) and fuel rate ($\dot{m}_f$). This is the sort of comparison discussed in connection with Fig. 10. First, at constant fuel rate, indicated power (P_{ind}) is proportional to indicated thermal efficiency (η_{ind}). From Fig. 8, at a given equivalence ratio (ϕ), the LHR engine shows a gain in η_{ind} relative to the baseline engine that ranges from a negligible to a small amount. But more importantly, beyond one-third load, η_{ind} falls markedly with increasing ϕ for both engines. This is primarily a result of increasing

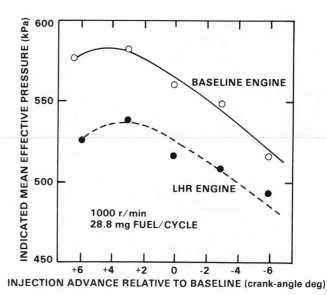

Fig. 12 - Effect of injection timing on indicated mean effective pressure at 1000 r/min, LHR and baseline engines at fixed fueling rate

combustion duration with increasing ϕ. Now, for the conditions of the simulated engine, at constant N and $\dot{m}_f$, ϕ varies inversely with volumetric efficiency (η_v). Thus at constant $\dot{m}_f$, a low η_v increases ϕ, which decreases η_{ind}. At the prescribed constant fuel rate, brake thermal efficiency (η_b) is represented by the following proportionality:

$$[\eta_b]_{\dot{m}_f} \propto (P_{ind} - P_m) = (\eta_{ind}\, \dot{m}_f\, Q_f - P_m) \qquad (1)$$

where fuel heating value (Q_f) is constant and motoring power (P_m) has been assumed a function of N only. It follows that at constant N and $\dot{m}_f$, the percentage loss in thermal efficiency, or inversely the increase in specific fuel consumption, is even greater on a brake basis than on an indicated basis as volumetric efficiency falls.

Volumetric efficiency is plotted against overall equivalence ratio for the two engines in Fig. 13. Superimposed dashed lines show that the loss in volumetric efficiency at constant fuel rate is even greater than at constant equivalence ratio. It seems clear that the comparatively poor BSFC of the naturally aspirated LHR engine is largely the result of the loss in volumetric efficiency.

Heat Loss - Heat lost from the gas in a diesel-engine cylinder is comprised of a convective component, which is decreased by lowering wall conductance, and a radiative component from incandescent soot particles, on which wall conductance has no direct effect. In Fig. 14 the calculated heat rejection from the cylinder gas between intake valve closing and exhaust valve opening is compared for the two engines at 1000 r/min. The effectiveness of the 80 percent reduction in conductance in

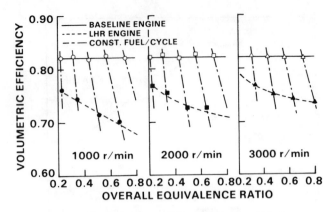

Fig. 13 - Volumetric efficiency (based on intake-manifold state) versus load at three speeds, LHR and baseline engines

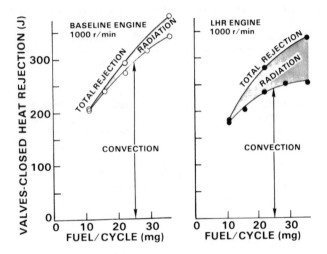

Fig. 14 - Convective and radiative heat rejection from LHR and baseline engines versus fueling rate at 1000 r/min

the LHR engine in decreasing convective loss is evident. However, both the resulting rise in flame temperature and the greater amount of soot formed increase the radiative component substantially, thus decreasing heat conservation from the level that might otherwise have been expected. Considerable uncertainty surrounds the validity of the soot model in the simulation. Hence, the impact of radiative heat loss needs to be watched closely as experiments proceed with the LHR diesel.

Another aspect of the heat-loss calculation deserving attention is the assumption of quasi-steady heat transfer, as characterized by wall temperatures that are invariant with crank angle. In reality the inner-surface temperature does fluctuate around some mean value during the cycle. Both the fluctuation amplitude and the mean vary from one location on the wall to another at a fixed operating point, creating a formidable problem for precise analysis. For a direct-injection engine, Wallace et al. have estimated a wall-temperature fluctuation of ± 3°C in a cast iron wall

and ± 10°C in a silicon nitride wall [6]. Working with a cast iron indirect-injection engine, Alkidas and Cole have measured a fluctuation of ± 4°C at a point in the prechamber and ± 6°C at a point in the main chamber [16]. It has been estimated that with silicon nitride walls, these could increase to as much as ± 27°C and ± 40°C respectively. These fluctuations are still small compared to the variation in cylinder-gas temperature during the cycle. Nevertheless, the walls would be below their average temperature during intake and above their average during combustion. The result of the former would be less degradation in volumetric efficiency for the LHR engine than shown in Fig. 13, but by an increment too small (perhaps 2 percentage points) to alter the qualitative conclusions based on the quasi-steady analysis.

EMISSIONS - From the earlier discussion of emission standards for passenger-car diesels, it appears essential that the LHR engine produce lower NOx emissions than current production diesel cars to meet future emission standards. The calculated emission indices of NOx (EINOx) are compared for the baseline and LHR engines as functions of overall equivalence ratio in Fig. 15 for operation at 1000 r/min. The approximate doubling of the EINOx in the LHR engine is a disappointment but not a surprise. Plee et al. have correlated EINOx with stoichiometric adiabatic flame temperature for indirect-injection diesels by means of an Arrhenius expression [21,22]. Because of the higher flame temperature in the LHR engine, application of their expression also projects an increase in NOx emission with heat conservation.

On the 13-mode heavy-duty test, the TACOM/ Cummins LHR diesel shows only a 12 to 23 percent increase in NOx compared to the cooled baseline engine at fixed injection timing [23]. This increase can be nullified at the expense of a small penalty in fuel consumption by retarding injection timing. However, at this

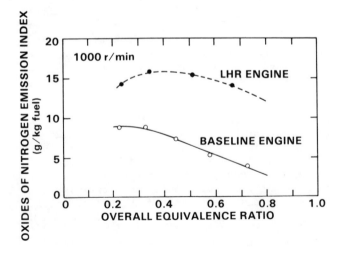

Fig. 15 - Emission index of oxides of nitrogen for LHR and baseline engines versus equivalence ratio at 1000 r/min

retarded equal-NOx timing, the fuel consumption with LHR was still less than for the baseline engine.

It has been conjectured that the NOx emission of the TACOM/Cummins engine is less sensitive to heat conservation than the indirect-injection engine that is the subject of this study because the former engine is of the quiescent-chamber direct-injection type. In such an engine the kinetic energy of the high-pressure fuel jets, rather than intense air motion, is relied upon to produce the necessary mixing of fuel and air. As a consequence of the comparative quiescence of the air during compression, it has been suggested that heat transfer from the hot cylinder walls to the fresh air charge is limited to a thermal boundary layer of air adjacent to the walls and need not involve the core air participating in combustion to an appreciable extent. The resulting failure of the hot cylinder walls to effect a marked increase in flame temperature in the core could reduce the sensitivity of the NOx emission to heat conservation. If this speculation is correct, it cannot be carried over to the indirect-injection engine because in that construction it is the mixture motion associated with flow into and out from the prechamber that is responsible for the essential mixing function, with the kinetic energy of the fuel jet playing only a minor role [24]. In such an engine, significant heat exchange between the combustion chamber walls and the combustion air is inevitable.

The calculated emission indices of particulate carbon (EIC) for the baseline and LHR engines are plotted against overall equivalence ratio at 1000 r/min in Fig. 16. Although the EIC levels are comparable at light load, EIC is projected to increase faster with load in the LHR engine than in the baseline engine. This contrasts sharply with the particulate reduction of over 80 percent claimed for the TACOM/Cummins LHR diesel [23]. It is also contrary to what would be expected from the Arrhenius correlation of Plee et al. [21,22], whose experimental findings would lead one to expect the higher flame temperature of the LHR

engine to decrease EIC. The net production of soot is the result of two competing mechanisms, however. The first, pyrolytic formation of soot occurring early in the combustion process, is fostered by high temperature. The second, subsequent destructive oxidation of soot particles, is also favored by high temperature. That Plee et al. found net soot production to fall with increasing flame temperature in conventional engines indicates that in their experiments, oxidation was dominant. There is no assurance that a radical change in combustion environment imposed by heat conservation cannot reverse the outcome of this competition between formation and oxidation. On the other hand, it is uncertain that the Khan expression used to model formation [14] and the Nagle and Strickland-Constable expression used to model oxidation [15] accurately represent in-cylinder soot mechanisms. The findings of this study underscore the need for a confirmed submodel of net soot production in diesel engines, and they also point to the need for careful monitoring of exhaust particulates from the LHR diesel as experimental developments proceed, until this issue is satisfactorily resolved.

SUPERCHARGING

Turning again to fuel economy, it was evident from the analysis of the naturally aspirated LHR engine that a major reason it showed no significant gain was the penalty it suffered in volumetric efficiency, the result of decreased cylinder-air density from heating by the cylinder walls during intake. One approach to overcoming this defect is to supercharge the engine. To analyze the results of this cycle modification necessitates modeling the total engine system to define the interaction between the LHR engine and the supercharging compressor it drives. This requires simultaneous satisfaction of the following three conditions:

1. Compressor rotational speed equals the product of engine speed and compressor-drive speed ratio.

2. The mass airflow rates through the compressor and into the engine are equal.

3. The pressure rise through the compressor equals the pressure drop across the engine (assuming ambient pressure at the compressor inlet and engine discharge).

Then the engine brake power equals the algebraic sum of indicated power, engine friction power, and the power required to drive the compressor.

The chart in Fig. 17 is helpful in illustrating how the three conditions specified above are satisfied. The upper half of the chart characterizes certain aspects of a positive-displacement supercharging compressor of the internal-compression, or screw, type. Compressor performance is normally represented on a map of compressor pressure ratio versus mass flow rate containing lines of constant speed, with contours of constant adiabatic efficiency

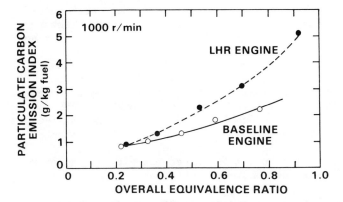

Fig. 16 - Emission index of particulate carbon for LHR and baseline engines versus equivalence ratio at 1000 r/min

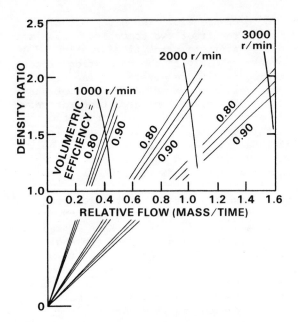

Fig. 17 - Matching chart for engine-driven positive-displacement supercharger

superimposed. When the engine inhales compressor discharge air, however, it is air density rather than pressure that is of major interest. The density ratio across the compressor is related to the pressure ratio by

$$\frac{\rho_2}{\rho_1} = \frac{p_2}{p_1}\left(\frac{\eta_c}{\eta_c + Y}\right) \tag{2}$$

where η_c is the compressor adiabatic efficiency and Y is a function of the compressor pressure ratio and the ratio of specific heats (k), given by

$$Y = \left(\frac{p_2}{p_1}\right)^{\frac{k-1}{k}} - 1 \tag{3}$$

Using these expressions, compressor density ratio can be calculated from the traditional compressor map and plotted against flow for constant engine speed (given the compressor-drive speed ratio), as shown by the three nearly vertical curves of Fig. 17. The near verticality of these constant-speed curves reflects the characteristics of a constant-displacement compressor.

While these constant-speed curves indicate the flow capabilities of the isolated compressor, the four-stroke engine will swallow compressor mass flow only at a rate ($\dot{m}_a$) corresponding to

$$\dot{m}_a = \eta_v\left(\frac{\rho_2}{\rho_1}\right)\frac{N}{2}\rho_1 \, G \, D \tag{4}$$

where compressor inlet density (ρ_1), drive speed

ratio (G) and engine displacement (D) are constant. At each engine speed (N) this expression defines the triad of lines in Fig. 17 radiating from the imaginary zero-zero origin for volumetric efficiencies (η_v, based on intake manifold density) of 0.80, 0.85 and 0.90.

Each intersection of a ray of constant volumetric efficiency with its corresponding speed line defines a possible operating point. By feeding the compressor discharge pressure and temperature corresponding to each intersection into the engine cylinder model in order to determine the corresponding volumetric efficiency of the engine, it is then possible to establish which is the correct intersection for each engine speed and fuel rate by matching volumetric efficiencies. The engine cylinder model provides the associated indicated and pumping power. The power requirement of the compressor is given by

$$P_c = \dot{m}_a \frac{R \, k}{k-1}\frac{T_1 \, Y}{\eta_c} \tag{5}$$

where T_1 = compressor inlet absolute temperature
R = gas constant

Because of the increased cylinder mean effective pressures anticipated with supercharging, the previously given load-independent motoring mean effective pressures taken to represent mechanical friction were adjusted slightly with changing load according to the recommendation of Taylor and Taylor [20]. It developed that this modification was too small to have a significant effect on the conclusions reached from this study.

Applying this methodology, as matched in this example the supercharger is called upon to supply a pressure ratio of from 2.45 (light load) to 2.5 at 3000 r/min, 2.35 to 2.5 at 2000 r/min, and 1.85 to 2.15 at 1000 r/min.

The results at 2000 r/min are examined more closely in Fig. 18. The second-highest dashed curve shows indicated power as a function of fuel rate. Because cylinder pressure is higher during the intake stroke than during the exhaust stroke in a supercharged engine, the pumping work is positive. The top dashed curve results from adding pumping power to indicated power. That positive pumping comes at the expense of the shaft work put into the compressor, represented by the decrement (-C) subtracted from the top curve. The placement of the resulting curve reveals that more work was required to compress the air in the supercharging compressor than was recouped in the cylinder during the intake and exhaust strokes. Subtracting engine friction power from this curve provides the solid curve of brake power.

The dotted curve of brake power for the naturally aspirated baseline engine at 2000

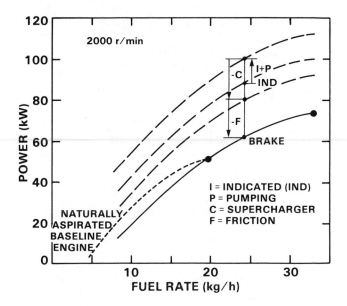

Fig. 18 - Power breakdown for supercharged LHR engine compared to brake power of baseline engine at 2000 r/min

r/min is added to Fig. 18 for comparative purposes. It is seen that the supercharged LHR engine offers 43 percent more brake power than the baseline at this speed, but at the expense of an increased fuel rate. Recalling that rays radiating from the origin of Fig. 18 are lines of constant specific fuel consumption, it is evident that the only condition at which the supercharged LHR engine can match the brake fuel economy of the baseline engine at this speed is at the fuel rate corresponding to full load of the baseline engine. At all other power levels the brake fuel economy of the supercharged baseline engine is inferior. It can be deduced from Fig. 18 that this occurs largely because only a portion of the shaft work put into the compressor is recovered by the piston during the gas exchange process.

The comparative fuel economy of the supercharged LHR engine can be improved by scaling engine displacement down to match the full-load power of the baseline engine. If this is done at 2000 r/min, the 5.7-L naturally aspirated baseline engine is matched by a 4.0-L supercharged LHR engine.

To provide some perspective of this approach, brake power is plotted against fuel rate for several engines in Fig. 19. The curves for engines A, the 5.7-L baseline engine, and B, the supercharged LHR version of the same engine, are repeated from Fig. 18. Curve C results from scaling the 5.7-L supercharged LHR engine down to 4.0 L. Although this produces an engine that consumes less fuel than the 5.7-L supercharged engine at power levels below 40 kW, it still consumes more fuel than the baseline engine over the entire load range.

Engine D in Fig. 19 results from declutching the supercharger of the 4.0-L LHR engine below 60 percent of full-load power. Above 31 kW, which is the power level at which the

unsupercharged 4.0-L LHR engine would reach the full-load fuel-air ratio of 0.0555, the supercharger is clutched in to provide the boost needed to match the full-load capability of the 5.7-L baseline engine. This trajectory is represented by the dotted line leading to the point that marks the full-load point for engine C. In principle, then, engine D is seen to offer better fuel economy than the baseline engine below 27 kW, but not above, and to match the full-load power of the baseline engine, but at a higher fuel rate. The practical problem of modulating power between 31 kW and 52 kW (full load) fall beyond the scope of this study.

It should be recognized that similarly equipping a 4.0-L version of the baseline engine with a declutchable supercharger would also improve its fuel economy.

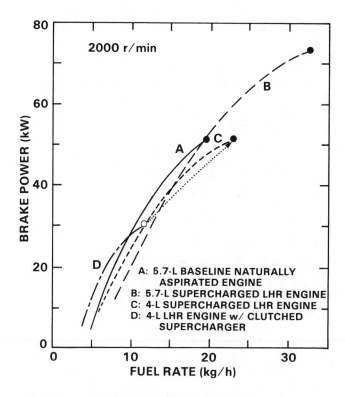

Fig. 19 - Brake power at 2000 r/min versus fueling rate for baseline engine, supercharged LHR engine, and downsized supercharged LHR engine

BOTTOMING CYCLE

The analysis of the naturally aspirated LHR engine indicated that most of the energy conserved was discharged from the engine along with the exhaust gas. One way of recovering some of this energy is by operating a bottoming cycle from the heat available in the exhaust stream. Working under contract to the Department of Energy, Thermo Electron (TECO) has demonstrated a Rankine cycle using an organic working fluid on a heavy-duty truck powered by a conventional turbocharged direct-injection diesel [25]. The bottoming cycle has also been considered for use on a heavy-duty LHR engine [5]. In the present study a preliminary assessment of the addition of a Rankine bottoming cycle to the naturally aspirated indirect-injection LHR diesel was made to determine its potential benefits relative to other methods for utilizing the heat conserved in such an engine.

The TECO bottoming cycle uses Fluorinol, a mixture of trifluoroethanol and water, as a working fluid. Although probably a good choice for the TECO application, it has some obvious shortcomings for use in a passenger car powered by a LHR diesel. First, it is expensive. Second, questions have been raised about the potential toxicity of working fluid leaked from the system. Third, its maximum operating temperature is limited by chemical stability, which could be a drawback to its efficient use with the increased exhaust gas temperatures typical of the LHR engine.

All three of these shortcomings are overcome by replacing Fluorinol with water. On the other hand, water is probably not a suitable working fluid in a production system because of its high freezing point. Known antifreezes suffer from the same problem of thermal instability as Fluorinol. Faced with this dilemma, water is felt to be an acceptable working-fluid choice in this study, the chief purpose of which is merely to gain insight into the characteristics to be expected if a Rankine bottoming cycle is appended to LHR passenger-car diesel.

Four problems associated with application of the Rankine bottoming cycle are independent of the working fluid selected. First, the bottoming-cycle hardware occupies considerable space and adds mass to the power plant. Although these handicaps may be manageable in a heavy-duty truck, in a modern passenger car they would pose quite a problem. Second, the complexity introduced by adding a second cycle normally entails a significant additional cost, which needs to be evaluated carefully for each intended application. Third, because of soot in diesel exhaust, the gas-side surfaces of the vapor generator quickly become coated with deposits that impair the transfer of heat into the working fluid. Cleaning techniques have met with some success, but this problem still poses a serious impediment to acceptance of a bottoming cycle in passenger-car service. Finally, the ability of the Rankine bottoming cycle to respond to the highly transient operation experienced in passenger-car operation has to be a concern.

A simple schematic of the Rankine-cycle arrangement is shown in Fig. 20. A turbine expander was chosen. The turbine pressure ratio proved to be sufficiently high that a multistage unit would be needed, with the first-stage nozzle being choked, i.e., operating with a sonic throat velocity, at all engine speeds and loads. This turbine, although not designed in detail, would clearly be

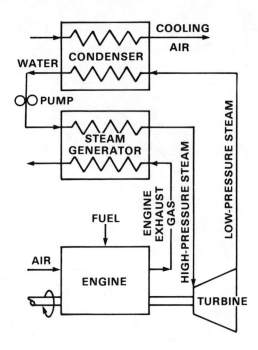

Fig. 20 - Schematic of engine with Rankine bottoming cycle

a small unit running at very high speeds and would require a sophisticated reduction gear to join it to the crankshaft.

The mass flow rate and temperature of the engine exhaust entering the steam generator were calculated as functions of engine speed and fuel-air ratio using the engine cylinder model. Exhaust temperature ranged from about 250°C at light load to 850°C at full load, with the temperature being 50-100°C lower at 1000 r/min than at 3000 r/min. In the interest of showing the Rankine cycle in a favorable light, an adiabatic exhaust system was considered to join the engine cylinders to the steam generator, and the pressure drop on the gas side of the steam generator was assumed negligible.

The rate of heat rejection to the condenser of the Rankine cycle was less than the rate of heat rejection to the radiator of the baseline engine at 1000 r/min. At 3000 r/min the reverse was true. On this basis it was reasoned that the power requirement of the condenser fan on the uncooled LHR engine would approximately balance the power requirement of the radiator fan on the baseline engine. Consequently, fan power requirements were excluded from the analysis of both engines.

In the Rankine-cycle steam loop, the condenser-discharge state point was fixed at a saturated condition of 105°C and 118.5 kPa. Steam-generator discharge temperature was limited to a maximum of 650°C. The pressure leaving the steam generator and entering the turbine was chosen to prevent excessive condensation in the turbine discharge stream in order to avoid turbine blade erosion.

The rate at which steam is circulated around the steam loop (m_s) is an independent variable. Combining the definition of steam generator efficiency (η_g) with a heat balance in the steam generator,

$$\dot{m}_s = \frac{\dot{m}_x \, \frac{R \, \gamma}{\gamma - 1} \, \eta_g \, (T_{x,in} - T_{g,in})}{h_{t,in} - h_{g,in}} \qquad (6)$$

where $\dot{m}_x$ = engine exhaust mass flow rate

$T_{x,in}$ = engine exhaust temperature entering steam generator

$T_{g,in}$ = water temperature entering steam generator

$h_{t,in}$ = steam enthalpy entering turbine

$h_{g,in}$ = water enthalpy entering steam generator

R = gas constant for engine exhaust

γ = ratio of specific heats for engine exhaust

In this expression, $T_{g,in}$ and $h_{s,in}$ (both fixed by the condenser), R, and γ are all either constant or reasonably constant. $T_{x,in}$ is determined by the engine, increasing substantially with fuel-air ratio and to a much smaller degree with speed. η_g, estimated from heat transfer theory, is nearly independent of fuel-air ratio at a given speed but increases with decreasing speed from 0.72 to 0.80 to 0.90 for 3000, 2000 and 1000 r/min, respectively. On a percentage basis, this change is small compared to the change in $\dot{m}_x$, which varies by a factor of three over this speed range. What this expression shows, then, is that if maximum Rankine-cycle temperature (as characterized by $h_{t,in}$) is to be kept high in order to preserve high efficiency in the bottoming cycle as engine speed is varied at a given fuel-air ratio, then steam mass flow rate ($\dot{m}_s$) must be decreased as engine speed (and with it $\dot{m}_x$ and $T_{x,in}$) is lowered.

Mass flow rate through the choked turbine nozzle is represented by

$$\dot{m}_s \propto \frac{P_{t,in} \, A_n}{\sqrt{T_{t,in}}} \qquad (7)$$

where $P_{t,in}$ = turbine inlet absolute pressure

A_n = effective nozzle flow area

At fixed fuel-air ratio, if $\dot{m}_s$ is to fall with decreasing engine speed without significantly changing $T_{t,in}$, then either turbine inlet pressure or nozzle area must be reduced. Lowering $P_{t,in}$ adversely affects cycle efficiency by decreasing the pressure ratio across the turbine to the fixed condenser pressure. It is concluded, then, that to maximize the contribution of the Rankine bottoming cycle in an application covering a broad range of engine speed, A_n must decrease with falling engine speed. This calls for a variable-admission turbine. Nozzle area is made a function of engine speed only. On a relative basis, the nozzle area required in this design ranges from

100 percent at 3000 engine r/min to 65 and 24 percent at 2000 and 1000 r/min, respectively. The resulting variations in steam flow rate and in turbine inlet pressure and temperature with fuel-air ratio are shown in Fig. 21.

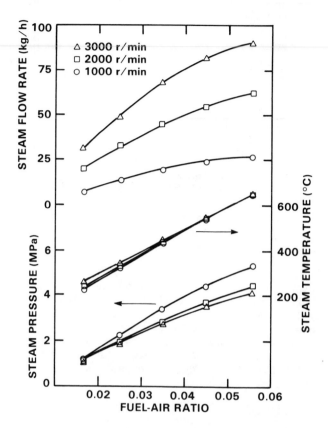

Fig. 21 - Pressure, temperature and steam flow rate for Rankine bottoming cycle

The efficiency of a turbine is a function of the ratio of tangential rotor-blade speed to the velocity equivalent of the isentropic enthalpy drop across the turbine. A typical parabolic variation of turbine efficiency is plotted in Fig. 22. Because the turbine speed remains proportional to engine speed while the velocity equivalent of the enthalpy drop does not, the turbine cannot operate at its maximum efficiency most of the time. The operating range at each engine speed is indicated in Fig. 22, with full load at the left of each range and light load at the right. The fact that the turbine efficiency approaches zero at 3000 r/min and no load is inconsequential because the engine seldom operates there.

Although, as previously shown, the naturally aspirated LHR engine offered no fuel-economy advantage relative to the baseline engine over the important part of the operating range, the addition of the Rankine cycle to the LHR engine does provide a consistently superior fuel economy. The LHR/Rankine engine also delivers more full-load power at 3000 r/min. After scaling the LHR/Rankine engine down 15 percent to limit its

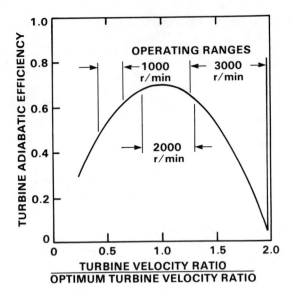

Fig. 22 - Steam turbine efficiency versus relative turbine velocity ratio

output at 3000 r/min to that of the baseline engine, the LHR engine with the bottoming cycle shows about a 15-percent lower fuel consumption than the baseline over nearly the entire speed/load range. This compares with a real-world 13 percent measured for the TECO unit behind a conventional diesel in heavy-duty truck service [25].

TURBODIESEL

It has been shown that supercharging with an engine-driven blower compensates for the deterioration in volumetric efficiency associated with the naturally aspirated LHR diesel, but not in a way that reaps large benefits in fuel economy. This approach fails to capitalize on the increased energy content of the exhaust gas when cylinder heat conservation is introduced.

It has also been shown that making use of this exhaust heat in a bottoming cycle can produce an attractive gain in fuel economy. However, concerns about space requirements, added mass and cost, and fouling of heat-transfer surfaces with soot raise serious questions about the applicability of the bottoming cycle to the passenger car.

The turbodiesel, which involves combining the diesel engine with turbomachinery, is a third approach. It offers the possibility of making use of the extra exhaust energy while concurrently compensating for deteriorated volumetric efficiency. Projections of fuel economy for the turbodiesel have been avoided here because they are extremely sensitive to predicted cylinder exhaust-gas temperatures under boosted conditions and would thus extend use of the phenomenological cylinder model into regions for which it has not yet been validated. Through simple analysis, however,

it is possible to gain some insights into the LHR turbodiesel.

In considering the addition of turbo-machinery to an engine with a boosted intake manifold pressure, one of the first questions to ponder is how the energy-extraction capability of an exhaust turbine that imposes a back pressure p_x in the exhaust manifold compares with the ability of the piston to extract work at the same exhaust manifold pressure by simply relying on the positive pumping work generated during the gas-exchange process. This question is conveniently addressed by comparing idealized cycles for two different engine arrangements: one in which the cylinders are neutrally supercharged, i.e., in which the boost pressure in the intake manifold (p_i) and the turbine-induced back pressure in the exhaust manifold (p_x) are equal and greater than atmospheric pressure (p_a), and one in which the engine is boosted to the same p_i but releases the cylinder contents to an exhaust manifold pressure $p_x = p_a$.

Expressions for the power available from an ideal, isentropic turbine ($P_{t,id}$) and from an ideal pumping loop ($P_{p,id}$) are given in Appendix C. For the ideal pumping loop, let $p_i/p_x = \Pi = p_i/p_a$. For the engine employing the exhaust turbine, $p_x/p_a = p_i/p_a = \Pi$. Then

$$\frac{P_{t,id}}{P_{p,id}} = \frac{\left[\eta_v \, (1+F) \, \dfrac{\gamma}{\gamma-1} \right] \dfrac{T_x}{T_i} \, \Pi \left[1 - \Pi^{\frac{1-\gamma}{\gamma}} \right]}{\Pi - 1} \qquad (8)$$

Considering the quantities in the first set of brackets, the volumetric efficiency based on intake manifold conditions (η_v), one plus the fuel-air ratio (F), and the ratio of specific heats in the exhaust (γ) normally vary only over a limited range. On the other hand, the ratio of cylinder exhaust gas temperature (T_x) to intake manifold temperature (T_i) can easily vary from 2 at light load to 4 at full load. After substituting typical values into Eq. 8, the range of ratios of turbine power to pumping power is presented as a function of pressure ratio (Π) in Fig. 23. It is seen that in principle, the turbine is two to four times as effective in extracting energy as the piston.

But Eq. 8 and Fig. 23 are for ideal conditions. The real turbine does not provide isentropic expansion, as assumed. A maximum adiabatic efficiency in the vicinity of 0.80 is reasonable for the small turbines that match automotive engines, meaning that values on the ordinate of Fig. 23 need to be multiplied by 0.8. If the turbine is constrained to operate at a speed other than its optimum for the head available, e.g., by being geared to the crankshaft, then its efficiency can fall substantially below the maximum level. This was seen to occur in the Rankine-cycle turbine (Fig. 22). Considering the magnitude

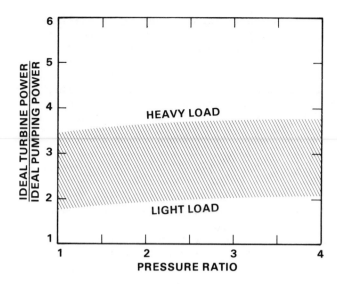

Fig. 23 - Ratio of output from ideal turbine to output from ideal positive pumping loop

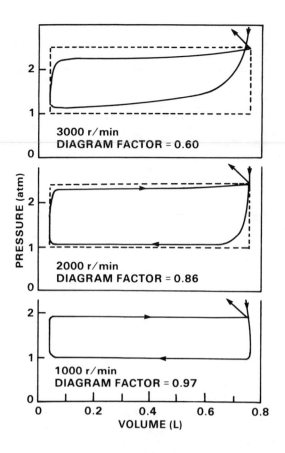

Fig. 24 - Calculated pressure-volume diagram for gas exchange process compared to ideal rectangular diagram

of the advantage of the ideal turbine in Fig. 23, however, the turbine efficiency would have to fall below 50 percent at light loads and low pressure ratios ($P_{t,id}/P_{p,id} \simeq 2$) and below 20 percent at heavy loads and high pressure ratios ($P_{t,id}/P_{p,id} \simeq 5$) in order for it to be a less efficient energy converter than the piston ideally is during the gas exchange process.

The piston does not necessarily follow its ideal model either. In Fig. 24 the calculated pressure-volume diagrams are shown at three speeds for the intake and exhaust strokes of the previously analyzed supercharged LHR engine at an intermediate load (fuel-air ratio 0.03). In the upper two diagrams the ideal rectangular pumping loop is superimposed. At 3000 r/min the throttling that occurs across the intake and exhaust valves causes significant departures from the perfect rectangle. The diagram factor, defined as the ratio of the actual pumping work resulting from integration of the pressure-volume traces of the intake and exhaust strokes to the area of the rectangular diagram, is only 0.60. This further contributes to the superiority of the turbine expander. At 2000 r/min, with more time available for filling and emptying the cylinder, the diagram factor rises to 0.86. At 1000 r/min the pressure-volume trace during gas exchange approximates the ideal rectangle so closely that the difference between them is almost indiscernible. The resulting diagram factor is 0.97.

From this discussion it can be reasoned that if the cylinder pressure at BDC beginning the exhaust stroke exceeds ambient as a result of boosting intake manifold pressure, to maximize efficiency it is more desirable to expand the exhaust through a turbine than to extract the energy with the piston. That is, boosting the engine to a given intake manifold pressure with a turbocharger is more efficient

than boosting it with an engine-driven supercharger, given the same compressor efficiency in both cases. Note, however, that this reasoning was based on use of a neutral turbocharger, i.e., one that provides the same pressure in the intake manifold as it requires in the exhaust manifold to produce that boost. That is actually a special case. In practice, the pressure difference across the cylinders of a turbocharged engine may be either positive ($p_i > p_x$) or negative ($p_i < p_x$).

Experimental data from a turbocharged conventional diesel engine of the indirect-injection type are plotted in Fig. 25. In the lower part of the figure, intake manifold pressure is plotted against fuel-air ratio for four engine speeds. Unlike what was found previously for the supercharged engine, the pressure in the intake manifold is a strong function of fuel-air ratio. The horizontal segment of the characteristic at 3000 r/min results from the use of a wastegate that bypasses part of the cylinder exhaust around the turbine in order to avoid excessive cylinder pressures, as well as to shape the full-load torque curve for automotive use [26].

At the top of Fig. 25, the pressure difference across the engine cylinders is plotted. Exhaust manifold pressure is seen to exceed intake manifold pressure under nearly

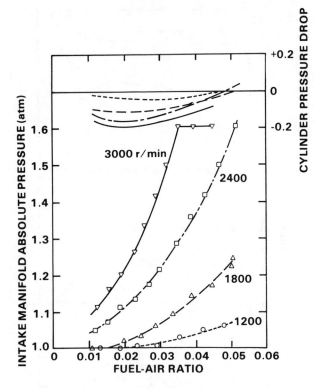

Fig. 25 - Intake manifold pressure versus fuel-air ratio for typical turbocharged diesel

efficiency, and the mechanical efficiency that represents the ratio of compressor to turbine shaft work

Y is given by Eq. 3, and Z is the following function of turbine pressure ratio (P_4/P_3) and the ratio of specific heats in the turbine:

$$Z = 1 - \left(\frac{p_4}{p_3}\right)^{\frac{\gamma-1}{\gamma}} \tag{10}$$

Assuming both compressor inlet and turbine exit pressures equal to ambient pressure, it is possible with this expression to relate the pressure difference across the cylinder to the other relevant variables, as is done for a compressor inlet temperature of 310 K in Fig. 26. It is apparent from this figure that at any given compressor pressure ratio, of which three are represented, an increase in either available turbine inlet temperature or turbocharger efficiency favors a more positive pressure difference across the cylinder. Although it appears from Fig. 26 that a movement in that direction would benefit the contemporary turbocharged passenger-car diesel, it is also seen that too much movement in that direction would result in an excessively large positive pumping loop that, from Fig. 23, fails to take full advantage of the pressure available in the exhaust manifold.

all conditions. Although the turbocharger increases the power available from a given engine displacement, the negative pumping loop associated with the pressure rise across the cylinders detracts from the efficiency with which this is accomplished. On the positive side, it should be recognized that mechanical-friction power normally rises at a slower rate than indicated power as boost is increased, thus contributing to brake thermal efficiency. The net result of these and other factors has been that the fuel economy advantage from turbocharging the passenger-car diesel has been modest at best. Clearly, eliminating the negative pumping loop would help.

The pressure difference across the cylinders is primarily a function of exhaust-gas temperature and turbocharger efficiency. Equating the work developed by the turbine to that absorbed by the compressor [27], it can be shown that

$$\frac{Y}{Z} = (1 + F)\ \frac{\gamma\ (k-1)}{k\ (\gamma-1)}\ \eta_{tc}\ \frac{T_3}{T_1} \tag{9}$$

where
F = fuel-air ratio
T_3 = turbine inlet absolute temperature
T_1 = compressor inlet absolute temperature
η_{tc} = turbocharger efficiency, the product of compressor adiabatic efficiency, turbine adiabatic

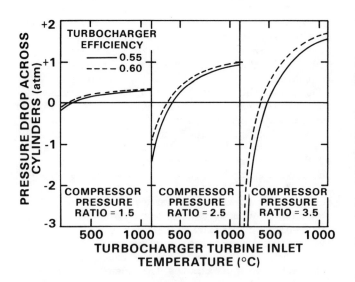

Fig. 26 - Effect of turbine inlet temperature and turbocharger efficiency on pressure drop available across cylinders of a turbocharged engine, no induction or exhaust losses

Because the LHR approach has been shown to offer increased cylinder exhaust-gas temperature, corresponding to rightward movement in Fig. 26, the turbocompound arrangement of Fig. 27 has been deemed particularly attractive for

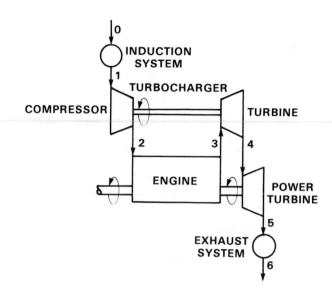

Fig. 27 - Schematic of turbocompound diesel showing station numbers

$$\frac{p_3}{p_4} = \frac{p_2/p_1}{1 + 0.08\sqrt{\dfrac{p_2}{p_1} - 1}} \qquad (13)$$

Then applying Eq. 9 and 13 to a turbocharged engine with a compressor inlet temperature of 310 K and a turbocharger efficiency of 0.55, the turbine inlet temperature required for the neutral condition of equal exhaust- and intake-manifold pressures is represented by the cross-hatched band in Fig. 28. The lower extremity of the band corresponds to full-load operation, the upper to operation at a light-load fuel-air ratio of 0.015.

the LHR diesel. In this arrangement a power turbine geared to the engine crankshaft is located downstream from the turbocharger turbine to extract exhaust energy made available by the elevated exhaust-manifold pressure that would otherwise go into a positive pumping loop. Although this layout has special merit for the heavy-duty application, which involves operation near full load a good share of the time, it needs special consideration for the light-duty passenger car.

To illustrate, it is seen from Fig. 27 that combining the pressure ratios of the individual components leads to

$$\frac{p_1}{p_0} \times \frac{p_2}{p_1} \times \frac{p_3}{p_2} \times \frac{p_4}{p_3} \times \frac{p_5}{p_4} \times \frac{p_6}{p_5} = 1 \qquad (11)$$

because $p_0 = p_6 =$ ambient pressure.

For the special case of a neutrally turbocharged engine, p_3 equals p_2, and because of deletion of the power turbine, p_4 equals p_5. Then for this special case the pressure ratio across the turbocharger turbine is related to the compressor pressure ratio by

$$\frac{p_3}{p_4} = \left(\frac{p_1}{p_0} \times \frac{p_6}{p_5}\right) \frac{p_2}{p_1} \qquad (12)$$

where the pressure ratios within the parentheses reflect the parasitic pressure losses across the induction and exhaust system that must be taken into account when Eq. 9 is applied to an engine in the as-installed condition. Adopting an approximate, experimentally based model of the parasitic losses for illustrative purposes, it is assumed that

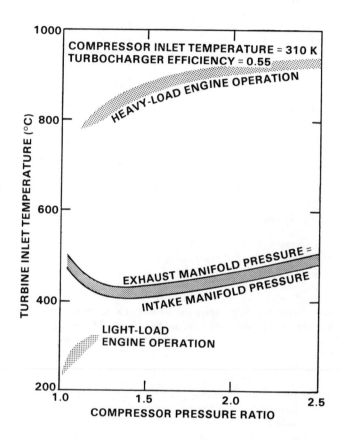

Fig. 28 - Effect of engine load on ability to avoid a negative pumping loop in a turbocharged diesel

Also plotted in the same figure are the ranges of exhaust-gas temperatures one might expect from the LHR cylinder at both heavy and light engine loads. At heavy load the available turbine inlet temperature exceeds that required for a neutral turbocharger by a wide margin, indicating that the engine would benefit from addition of a power turbine. This reflects the situation confronting the heavy-duty diesel. At light load, however, the cylinder exhaust-gas temperature is insufficient to achieve that condition, creating a

107

negative pumping loop. In fact, in this example the condition of a pressure rise existing across the cylinders persists throughout the typical road-load range of engine operation. This raises the question of whether the superior energy-conversion ability of the turbine expander, relative to the performance of the piston during gas exchange, is enough to overcome the negative pumping work it promotes at light load. The appropriateness of this question is underscored by the experimental findings of Yoshimitsu et al. [28], who found that at less than half of rated speed in a water-cooled turbocompound engine, the power turbine was ineffective below one-third load and under some conditions actually consumed a slight amount of power from the crankshaft.

To consider the factors influencing this situation, calculations were done for a turbocompound engine of the type illustrated in Fig. 27 that is turbocharged to part-load compressor pressure ratios of 1.5 and 1.2 respectively. With parasitic pressure losses estimated from Eq. 13, the turbomachinery was designed for the neutral condition at the following design point:

Engine speed	3000 r/min
Fuel-air ratio	0.0555
Compressor pressure ratio	2.5
Compressor inlet temperature	310 K
Cylinder exhaust temperature	1220 K
Turbocharger efficiency	0.55
Compressor efficiency	0.75
Turbine efficiency	0.80
Power turbine efficiency	0.80

The method of Mallinson and Lewis [29] was used to establish the division of available pressure ratio between the two turbines at off-design conditions. To simplify this illustrative calculation, the following parameters were held constant: ratios of specific heats in the compressor and turbine, turbocharger component efficiencies, and part-load fuel-air ratio (= 0.015).

The interplay between power turbine output and pumping work at 2000 engine r/min is shown as a function of cylinder exhaust gas temperature in Fig. 29 for both compressor pressure ratios selected. The outputs from both elements are expressed in terms of mean effective pressure. Three points are of special interest from this plot.

First, perhaps surprisingly to some, the output from the power turbine actually increases as cylinder exhaust temperature falls. This happens because as that temperature decreases, a greater pressure ratio is required across the turbocharger turbine to maintain the specified compressor pressure ratio, as follows from Eq. 9. With fixed turbine geometries, the pressure ratio across the power turbine increases when the pressure ratio across the turbocharger turbine increases, and that pressure ratio

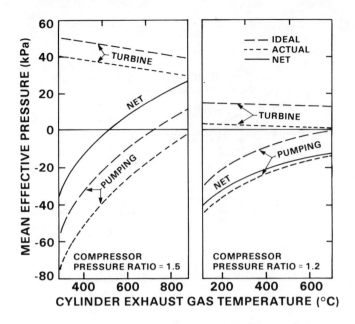

Fig. 29 - Output balance between the power turbine of a turbocompound diesel and the cylinder gas exchange process at two part-load compressor pressure ratios

increase is enough to outweigh the effect of the falling cylinder exhaust gas temperature. But that increase in power turbine output does not come free. As the cylinder exhaust temperature falls, the engine cylinder is called upon to act as a pump during the gas exchange process in order to supply the increased pressure-ratio requirement of the two turbines. This results in the increasingly negative pumping work shown in Fig. 29 as temperature falls.

The second point from Fig. 29 is that power turbine efficiency can be severely depreciated by the fact that the power turbine is geared to the crankshaft. As can be deduced from comparison of the ideal and actual turbine output curves, at the 1.5 pressure ratio the constraint of crankshaft speed is minimal. The isentropic head across the turbine has decreased from its design-point level in approximate proportion to the one-third reduction in engine speed from 3000 (design point) to 2000 (Fig. 29) r/min, so that turbine efficiency remains near the peak value of 0.80. The flatness of the typical turbine efficiency curve near its peak, as evident from Fig. 22, helps here. At the 1.2 pressure ratio, however, the turbine head has fallen without a commensurate change in speed. Now the turbine is forced to operate well beyond the optimum velocity ratio, as previously discussed in connection with the Rankine cycle and Fig. 22. Turbine efficiency falls below 0.20, explaining the proportionately wide spread between the ideal and actual turbine output curves at the lower compressor pressure ratio.

The third point follows from the solid net-output curves in Fig. 29 that result from adding actual outputs from the power turbine

and the cylinder gas exchange. The virtue of
the increased cylinder exhaust gas temperature
from the LHR diesel in eliminating, or at least
minimizing, negative net output is evident.
For the conditions of this study, a minimum
temperature of 500°C is required at a compressor pressure ratio of 1.5 in order for the
power turbine to break even. At the 1.2 pressure ratio the turbine never compensates for
the negative pumping work. It is also noted
that the mean effective pressures involved
represent only a small fraction of engine rated
output, which would be on the order of 1000
kPa. This points out the need for careful
attention to the role of the power turbine in
any turbocompound engine used primarily at
light load, as in the passenger car. The
cost/benefit tradeoff of including a power
turbine deserves special attention in this
case. Of course, even if the power turbine is
eliminated, the turbocharger can still use to
good advantage the elevated cylinder exhaust
gas temperature provided by the LHR engine.

CONCLUSIONS

1. In a LHR (low-heat-rejection) diesel,
the conservation of heat normally rejected to
the coolant offers a reservoir of otherwise
often-wasted energy that is particularly attractive in the passenger-car application because
of the proportionately high coolant heat loss
at light loads. However, the feasibility of
recovering a large fraction of this energy in
the basic engine itself, without appending
additional devices, is severely restricted by
the laws of thermodynamics.

2. Most of the energy conserved in the
LHR diesel escapes with the exhaust gas in the
form of increased gas temperature.

3. Incorporating LHR concepts in a
naturally aspirated diesel is unlikely to
improve fuel economy significantly because (a)
the increased temperature of the walls confining the cylinder charge depreciates volumetric efficiency, requiring additional piston
displacement, and with it greater friction, to
maintain performance, and (b) the simple
naturally aspirated engine fails to use the
increased energy content of the exhaust gas.

4. Simply adding an engine-driven supercharger to the LHR diesel to compensate for its
depreciated volumetric efficiency did not
improve fuel economy in the example considered
because (a) recovery of supercharger input work
by the piston during the gas exchange process
is poor, and (b) this arrangement is unable to
use the increased energy content of the exhaust
gas.

5. Declutching the supercharger in a
downsized supercharged LHR diesel having a
full-load power potential equal to that of the
conventional naturally aspirated baseline
engine offered some fuel economy advantage at
light load. The problem of modulating power
between unsupercharged and supercharged operation is not addressed in this study. Applying

this same technique to a downsized version of
the baseline engine would also improve its fuel
economy.

6. Adding a water-based Rankine bottoming
cycle to the naturally aspirated LHR engine is
projected to offer an upper-limit fuel economy
improvement, relative to the conventional
baseline engine of equal power, of about 15
percent. The bottoming cycle is considered
impractical for a passenger car, however,
because of such matters as space requirement,
mass, cost, fouling of the steam generator with
exhaust soot, and control during the highly
transient driving characterizing passenger-car
operation.

7. Combining the LHR diesel with turbomachinery offers the most attractive route to
improved fuel economy through heat conservation.
This option both (a) facilitates inlet-pressure
boosting to compensate for depreciated volumetric efficiency, and (b) capitalizes on the
increased energy content of the exhaust gas.

8. The turbocompound version of the LHR
diesel favored for heavy-duty use, which makes
use of both a turbocharger and an in-series
power turbine geared to the crankshaft, deserves
scrutiny before transferring it to passenger-
car use. At the low engine speeds and light
loads typifying this application, the power
turbine contributes little power and may even
become a parasitic load. Increased turbomachinery efficiencies and increased cylinder
exhaust gas temperature both help to alleviate
this situation. If the power turbine is
eliminated as the result of a cost/benefit
tradeoff, the turbocharger is still able to use
the higher cylinder exhaust gas temperature
beneficially.

9. The combustion model for the indirect-
injection diesel used in this study projected a
substantial increase in the emission of oxides
of nitrogen for the LHR engine. This is in
keeping with established correlations between
NOx emission and adiabatic flame temperature
for this type of engine.

10. The combustion model used projects an
increase in production of particulate carbon
for the LHR engine. This tentative projection
is contrary to established correlations between
particulates and adiabatic flame temperature
for this type of engine, and to data published
for an LHR diesel of the direct-injection type.
The discrepancy reflects either a failure of
the current model, which is based on published
expressions for soot formation and carbon
oxidation, or a switch in the indirect-injection LHR diesel from domination of the net soot
production process by formation rather than
oxidation as flame temperature is increased.

11. The combustion model used attributes
a significant component of the heat rejected to
the confining walls at high loads to radiation
from incandescent carbon particles. Decreasing
wall conductance for heat conservation has no
direct effect on this component. In view of
the uncertainties surrounding the soot model,
the impact of radiative heat loss on heat-

conservation efforts deserves close attention as experiments with the LHR diesel progress.

ACKNOWLEDGMENTS

The authors wish to express their gratitude to J. R. Mondt and A. C. Alkidas for helpful discussions on heat transfer matters, and to R. W. Talder for providing the experimental data on which Fig. 25 is based.

REFERENCES

1. G. Flynn and S. Timoney, "A Low Friction Unlubricated Silicon Carbide Diesel Engine," SAE Paper 830313, 1983.

2. H. A. Burley and T. L. Rosebrock, "Automotive Diesel Engines -- Fuel Composition vs. Particulates," SAE Transactions, Vol. 88, pp. 3112-3123 (Paper 790923), 1979.

3. O. A. Ludecke and D. L. Dimick, "Diesel Exhaust Particulate Control System Development," SAE Paper 830085, 1983.

4. R. Kamo and W. Bryzik, "Cummins-TARADCOM Adiabatic Turbocompound Engine Program," SAE Transactions, Vol. 90, pp. 263-274 (Paper 810070), 1981.

5. R. Kamo and W. Bryzik, "Adiabatic Turbocompound Engine Prediction," SAE Transactions, Vol. 87, pp. 213-223 (Paper 780068), 1978.

6. F. J. Wallace, R. J. B. Way and H. Vollmert, "Effect of Partial Suppression of Heat Loss to Coolant on the High Output Diesel Engine Cycle," SAE Paper 790823, 1979.

7. R. J. B. Way and F. J. Wallace, "Results of Matching Calculations for Turbo-charged and Turbocompound Engines with Reduced Heat Loss," SAE Paper 790824, 1979.

8. J. F. Tovell, "The Reduction of Heat Losses to the Diesel Engine Cooling System," SAE Paper 830316, 1983.

9. S. H. Mansouri, J. B. Heywood and R. Radhakrishnan, "Divided-Chamber Diesel Engine, Part I: A Cycle-Simulation Which Predicts Performance and Emissions," SAE Paper 820273, 1982.

10. R. T. Kort, S. H. Mansouri, J. B. Heywood and A. Ekchian, "Divided-Chamber Diesel Engine, Part II: Experimental Validation of a Predictive Cycle-Simulation and Heat Release Analysis," SAE Paper 820274, 1982.

11. X. Liu and D. B. Kittelson, "Total Cylinder Sampling from a Diesel Engine (Part II)," SAE Paper 820360, 1982.

12. C. A. Amann, D. L. Stivender, S. L. Plee and J. S. MacDonald, "Some Rudiments of Diesel Particulate Emissions," SAE Transactions, Vol. 89, pp. 1118-1147 (Paper 800251), 1980.

13. D. C. Siegla and G. W. Smith (Editors), Particulate Carbon Formation During Combustion, Plenum Press, New York, NY, 1981.

14. I. M. Khan, G. Greeves and D. M. Probert, "Prediction of Soot and Nitric Oxide Concentrations in Diesel Engine Exhaust," I. Mech. E. Symposium on Air Pollution Control in Transport Engines, pp. 205-217 (Paper C142/71), 1971.

15. J. Nagle and R. F. Strickland-Constable, "Oxidation of Carbon Between 1000-2000°C," Proceedings of the Fifth Carbon Conference, Vol. 1, pp. 154-164, 1962.

16. A. C. Alkidas and R. M. Cole, "Thermal Loading of the Cylinder Head of a Divided-Chamber Diesel Engine," SAE Paper 831325, 1983.

17. A. C. Alkidas and R. M. Cole, "The Distribution of Heat Rejection from a Single-Cylinder Divided-Chamber Diesel Engine," SAE Transactions, Vol. 90, pp. 2936-2948 (Paper 810959), 1981.

18. R. S. Radovanovic, K. F. Dufrane and R. Kamo, "Tribological Investigations for an Insulated Diesel Engine," SAE Paper 830319, 1983.

19. B. W. Millington and E. R. Hartles, "Frictional Losses in Diesel Engines," SAE Transactions, Vol. 77, pp. 2390-2410 (Paper 680590), 1968.

20. C. F. Taylor and E. S. Taylor, The Internal Combustion Engine, Second Edition, Chapter 12, International Textbook Company, Scranton, PA, 1966.

21. S. L. Plee, T. Ahmad, J. P. Myers and D. C. Siegla, "Effects of Flame Temperature and Air-Fuel Mixing on Emissions of Particulate Carbon from a Divided-Chamber Diesel Engine," Particulate Carbon Formation During Combustion, D. C. Siegla and G. W. Smith (Editors), Plenum Press, New York, NY, 1981, pp. 423-483.

22. S. L. Plee, T. Ahmad and J. P. Myers, "Flame Temperature Correlation for the Effects of Exhaust Gas Recirculation on Diesel Particulate and NOx Emissions," SAE Transactions, Vol. 90, pp. 3736-3753 (Paper 811195), 1981.

23. W. Bryzik and R. Kamo, "TACOM/Cummins Adiabatic Engine Program," SAE Paper 830314, 1983.

24. S. L. Plee and T. Ahmad, "Relative Roles of Premixed and Diffusion Burning in Diesel Combustion," SAE Paper 831733, 1983.

25. M. D. Koplow, L. D. DiNanno and F. A. DiBella, "Status Report on Diesel Organic Rankine Compound Engine for Long-Haul Trucks," Proceedings of the Twentieth Automotive Technology Development Contractors' Coordination Meeting, October 25-28, 1982, pp. 243-258.

26. C. A. Amann, "Why Not a New Engine?" SAE Transactions, Vol. 89, pp. 4561-4593 (Paper 801428), 1980.

27. R. W. Talder, J. D. Fleming, D. C. Siegla and C. A. Amann, "Evaluation of a Low-NOx Advanced Concept Diesel Engine for a Passenger Car," SAE Transactions, Vol. 87, pp. 1633-1652 (Paper 780343), 1978.

28. T. Yoshimitsu, K. Toyama, F. Sato and H. Yamaguchi, "Capabilities of Heat Insulated Diesel Engine," SAE Paper 820431, 1982.

29. D. H. Mallinson and W. G. Lewis, "The Part-load Performance of Various Gas-turbine Engine Schemes," Proceedings of the I. Mech. E., Vol. 159, pp. 198-219, 1948.

NOMENCLATURE

A	Effective area of heat rejection
A_n	Effective turbine-nozzle flow area
B	Intercept for Q_w vs. T_w (Eq. B-4)
C	Slope of Q_w vs. T_w (Eq. B-4)
D	Engine displacement
F	Fuel-air mass ratio
G	Gear ratio (supercharger speed/ engine speed)
h	Specific enthalpy
k	Ratio of specific heats for air
K	Ratio defined in Eq. B-5
$\dot{m}$	Mass flow rate
n	Polytropic exponent
N	Engine speed (rev/time)
p	Absolute pressure
$\bar{p}$	Mean effective pressure
P	Power
Q	Heat transferred
Q_f	Fueling heating value
R	Gas constant
T	Absolute temperature
U	Conductance
U*	Conductance ratio (Eq. B-3)
V	Volume
W	Work
Y	Pressure ratio function for compressor (Eq. 3)
Z	Pressure ratio function for turbine (Eq. 10)
γ	Ratio of specific heats for exhaust gas
Δ	Increment of
η	Efficiency
ν	Volumetric expansion ratio (Eq. A-7)
ρ	Density
ϕ	Fuel-air equivalence ratio
ψ	Work factor (Eq. A-7)

Subscripts

a	Air, ambient
b	Brake, baseline (Appendix B)
c	Compressor
cy	Cycle
f	Fuel
g	Steam generator

i	Intake manifold
id	Ideal
in	Into a component
ind	Indicated
m	Motoring
p	Pumping
s	Steam
t	Turbine
tc	Turbocharger
u	Underhood
v	Volumetric, based on intake-manifold state
w	Cylinder inner wall
x	Exhaust

Additional letter subscripts for cycle points used in Appendix A are defined in Fig. A-1.

Numbered station subscripts for the turbo-diesel are defined in Fig. 27.

APPENDIX A - WORK FACTOR

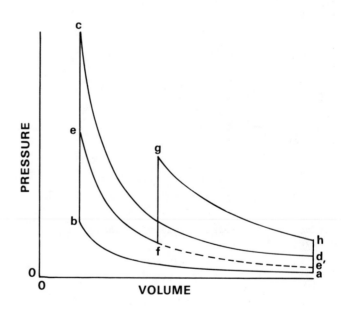

Fig. A-1

To assess the influence of the timing of heat addition/rejection on the production of indicated work in the cylinder gas, a dimensionless work factor (ψ) is introduced.

$$\psi = \frac{\Delta W}{\Delta Q} \tag{A-1}$$

where ΔQ = increment of heat addition/rejection for the cylinder gas
   ΔW = amount of ΔQ convertible to indicated work

In developing an expression for ψ, consider first the air-standard Otto cycle abcd of Fig. A-1. The cycle efficiency is

$$\eta_{cy} = 1 - \left(\frac{V_d}{V_e}\right)^{1-n} \tag{A-2}$$

$$= \frac{\text{Area } \overline{abcd}}{{}_b Q_c}$$

where   n = representative polytropic exponent

$_b Q_c$ = heat added in combustion, b to c

Of course, in a practical engine the combustion never occurs at constant volume as illustrated. However, an actual combustion event can be treated analytically by dividing the process into small increments of crank angle, each increment comprised of an isochoric heat addition followed by an adiabatic change of cylinder volume. Adopting that technique here, cycle abcd is modified such that half of the heat is added at TDC and the other half is added after the piston has descended through a third of its stroke.

The result is cycle abefgh, where

$${}_b Q_e = {}_f Q_g = \frac{{}_b Q_c}{2} \tag{A-3}$$

The efficiency of this cycle is

$$\eta_{cy} = \frac{\text{Area } \overline{abefgh}}{{}_b Q_e + {}_f Q_g}$$

$$= \frac{\overline{abee'} + \overline{e'fgh}}{{}_b Q_e} \tag{A-4}$$

But adapting the Otto cycle efficiency relationship of Eq. A-2,

$$\overline{abee'} = {}_b Q_e \left[1 - \left(\frac{V_e}{V_{e'}}\right)^{1-n} \right] \tag{A-5}$$

$$\overline{e'fgh} = {}_f Q_g \left[1 - \left(\frac{V_g}{V_h}\right)^{1-n} \right] \tag{A-6}$$

Thus each increment of heat added, or conversely of heat rejected, is seen to contribute to the cycle efficiency (indicated work) in proportion to its work factor

$$\psi = 1 - \frac{1}{v^{n-1}} \tag{A-7}$$

where v is the remaining volumetric expansion ratio from the point at which the increment of heat exchange occurs.

APPENDIX B

The heat-transfer and surface-temperature data base for this study came from the experiments of Alkidas and Cole [16], who measured these quantities in an indirect-injection diesel in which the coolant temperature was held at a constant 355 K. Considering that the majority of the heat loss in an indirect-injection diesel occurs across the main-chamber walls [17] and that the temperature differences across the walls of the cylinder and the prechamber were nearly identical at a given speed and fuel rate, the measured average main-chamber inner-wall temperature was used as the reference wall temperature T_{wb}. Then for a given engine operating condition the thermal resistance from the interior surface to the coolant can be represented by a baseline conductance

$$U_b = \frac{Q_{wb}}{A (T_{wb} - 355)} \tag{B-1}$$

where Q_{wb} = measured baseline heat rejection to coolant

A = effective area of heat rejection

T_{wb} = measured baseline inner-wall temperature (K)

Similarly, for the LHR engine,

$$U = \frac{Q_w}{A (T_w - T_u)} \tag{B-2}$$

where Q_w = unknown heat rejection to coolant

T_w = unknown inner wall temperature (K)

T_u = assigned underhood temperature (K)

The conductance ratio is then

$$U^* = \frac{U}{U_b} = \frac{Q_w (T_{wb} - 355)}{Q_{wb} (T_w - T_u)} \tag{B-3}$$

From a series of simulation runs for various wall temperatures at a fixed speed and fuel rate it was found that

$$Q_w \simeq B - C T_w \tag{B-4}$$

where B and C are empirical constants. Substituting Eq. B-4 in B-3 and rearranging, the unknown wall temperature is then

$$T_w = \frac{T_u + B K}{1 + C K} \tag{B-5}$$

where

$$K = \frac{(T_{wb} - 355)}{U^* Q_{wb}}$$

T_{wb} comes from the experimental results of the baseline engine, and Q_{wb}, B, and C from the simulation results, all evaluated at the same speed and fuel rate for which T_w is being sought. In this study the underhood temperature was fixed at 355 K. A value of 0.2 was selected for U^*.

APPENDIX C

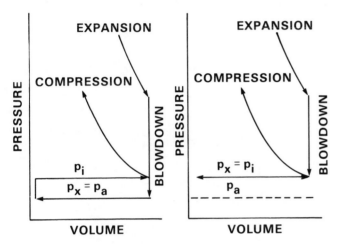

Fig. C-1

On the left in Fig. C-1, an idealized diagram of the gas-exchange process is shown for an engine that is boosted to intake pressure p_i and exhausts to a pressure p_x equal to ambient pressure p_a. The diagram on the right is for a boosted engine having equal exhaust and intake pressures, with the expansion to ambient pressure being completed in an exhaust turbine. The question addressed here is how the power available from the positive pumping loop of the engine diagrammed on the left compares with the power avaiable if the exhaust gas is instead expanded through a turbine as in the engine diagrammed on the right.

For the engine represented on the left, the power associated with the rectangular pumping loop is

$$P_{p,id} = \bar{p}_p \, D \, \frac{N}{2} = p_i \left[1 - \left(\frac{p_x}{p_i} \right) \right] D \, \frac{N}{2} \qquad (C-1)$$

For the engine illustrated on the right, the power available from isentropic expansion through a turbine from p_x to ambient pressure p_a is

$$P_{t,id} = \dot{m}_x \, \frac{\gamma \, R}{\gamma-1} \, T_x \left[1 - \left(\frac{p_a}{p_x} \right)^{\frac{\gamma-1}{\gamma}} \right]$$

$$= p_i \, \eta_v \, (1 + F) \, \frac{\gamma}{\gamma-1} \, \frac{T_x}{T_i} \, D \, \frac{N}{2} \left[1 - \left(\frac{p_a}{p_x} \right)^{\frac{\gamma-1}{\gamma}} \right]$$

$$(C-2)$$

120 page booklet.   Printed in U.S.A.